SPECTROSCOPIC REFERENCES
TO POLYATOMIC MOLECULES

IFI DATA BASE LIBRARY

COMPUTER TECHNOLOGY
Logic, Memory, and Microprocessors — A Bibliography
A. H. Agajanian

ACOUSTIC EMISSION
A Bibliography with Abstracts
Thomas F. Drouillard

MICROELECTRONIC PACKAGING
A Bibliography
A. H. Agajanian

SPEECH COMMUNICATION AND THEATER ARTS
A Classified Bibliography of Theses and Dissertations, 1973—1978
Merilyn D. Merenda and James W. Polichak

ATOMIC GAS LASER TRANSITION DATA
A Critical Evaluation
William Ralph Bennett, Jr.

BEHAVIORAL DEVELOPMENT OF NONHUMAN PRIMATES
An Abstracted Bibliography
Faren R. Akins, Gillian S. Mace, John W. Hubbard, and Dianna L. Akins

MASTER TABLES FOR ELECTROMAGNETIC DEPTH SOUNDING
INTERPRETATION
Rajni K. Verma

SPECTROSCOPIC REFERENCES TO POLYATOMIC MOLECULES
V. N. Verma

SPECTROSCOPIC REFERENCES TO POLYATOMIC MOLECULES

V. N. Verma

University of Khartoum
Khartoum, Sudan

IFI/PLENUM • NEW YORK-WASHINGTON-LONDON

ISBN-13: 978-1-4684-6119-0   e-ISBN-13: 978-1-4684-6117-6
DOI: 10.1007/978-1-4684-6117-6

© 1980 IFI/Plenum Data Company
Softcover reprint of the hardcover 1st edition 1980
A Division of Plenum Publishing Corporation
227 West 17th Street, New York, N.Y. 10011

INTRODUCTION

The rapid expansion of research activity in
all disciplines of science and technology
and the concomitant growth in the number of
scientific publications have caused a con-
siderable strain on library budgets. Even
well-endowed libraries find it too expensive
to collect all scientific journals. Research
workers therefore often find it difficult to
discover the extent and nature of work done
on a particular molecule. Moreover, liter-
ature searches, even in a well-equipped library, take a great deal of time and effort.

In recent years the need for a comprehensive
bibliography of the spectroscopy of polyatomic organic molecules, particularly the
derivatives of benzene, naphthalene, and
diazine, has been keenly felt. This volume
has therefore been prepared to provide easy
access to information on infrared, Raman,
absorption, emission, fluorescence, and
phosphorescence spectroscopy, and also re-
lated calculations. It covers about 900 organic ring compounds. These compounds are
arranged in alphabetic order; references are
given in chronological sequence for each
molecule, each characterized as to the tech-
nique and conditions covered.

Thus, much library time may be saved and, in
particular, even researchers who do not have
convenient access to a really good library
will be able to locate pertinent references
in a matter of seconds.

Acenaphthene ($C_{12}H_{10}$)

IR Spectrum — Proc. Roy. Soc. A195, 1 (1941)
    R.E. Richards and H.W. Thompson
Raman — Z. Phys. Chem. 195 (1950) — H.
    Luther and C. Reichel
Raman — J. Chim. Phys. 50, 250 (1954) —
    J.P. Mathieu, M. Ecollan and J.F.
    Ecollan
Raman — Acta Crystallogr. 10, 699 (1957) —
    H.W. Ehrlich
Fluorescence — J. Chim. Phys. 58, 204
    (1961) — L. Pesteil
Phosphorescence — J. Chim. Phys. 58, 204
    (1961) — L. Pesteil
IR (Crystal) — J. Chem. Phys. 39, 1942
    (1963) — L. Colombo
IR Spectrum — Opt. Spektrosk. 20, 1016
    (1966) — A.V. Nefedov and O.V.
    Fialkovskaya
Phosphorescence — Opt. Spektrosk. 22, 48
    (1967) — V.I. Mikhalenko and P.A.
    Teplyakov
Fluorescence — Opt. Spektrosk. 22, 48
    (1967) — V.I. Mikhalenko and P.A.
    Teplyakov
Vibrational Analysis — Spectrochim. Acta
    25A, 1815 (1969) — A.V. Bree, R.A.
    Kydd and T.N. Misra
Raman — Curr. Sci. 40, 251 (1971) — G.D.
    Baruah and S.R. Singh
IR Spectrum — Indian J. Pure Appl. Phys.
    10, 137 (1972) — G.D. Baruah, S.R.
    Singh, A.K. Chaudhary and D.K. Rai
UV Absorption in Vapour — Indian J. Pure
    Appl. Phys. 10, 305 (1972) — G.D.
    Baruah, P.S. Dube and S.R. Singh

Acenaphthene-D_{10} ($C_{12}D_{10}$)

Raman — Curr. Sci. 40, 251 (1971) — G.D.
    Baruah and S.R. Singh
Vibrational Analysis — Spectrochim. Acta
    28A, 627 (1972) — A.V. Bree and R.S.
    Williams
Vibrational Analysis — Indian J. Pure Appl.
    Phys. 10, 137 (1972) — G.D. Baruah,
    S.R. Singh, A.K. Chaudhary and D.K.
    Rai
UV Absorption in Vapour — Indian J. Pure
    Appl. Phys. 10, 305 (1972) — G.D.
    Baruah, P.S. Dube and S.R. Singh

Acetophenone ($C_6H_5CH_3CO$)

Vibrational Analysis — Spectrochim. Acta
    26A, 1097 (1970) — W.D. Mross and
    G. Zundel

Vibrational Analysis — C.R. Acad. Sci.
    (Paris)B228, 307 (1947) — M.C. Cherriey

Acetophenone-D_5 ($C_6D_5CH_3CO$)

Vibrational Analysis — Spectrochim. Acta
    26A, 1097 (1970) — W.D. Mross and
    G. Zundel

Acetophenone-D_8 ($C_6D_5CD_3CO$)

Vibrational Analysis — Spectrochim. Acta
    26A, 1097 (1970) — W.D. Mross and
    G. Zundel

p-Acetylaniline ($C_6H_4NH_2CH_3CO$)

IR Spectrum — Spectrochim. Acta 25A, 1423
    (1969) — L.K. Dyall

Acetylbenzene ($C_6H_5CH_3CO$)

See   Acetophenone

o-Acetylpyridine ($C_5H_5C_2H_2NO$)

Vibrational Analysis — Indian J. Phys. 51B,
    399 (1977) — K.C. Medhi

m-Acetylpyridine ($C_5H_5C_2H_2NO$)

Vibrational Analysis — Indian J. Phys. 51B,
    399 (1977) — K.C. Medhi

p-Acetylpyridine ($C_5H_5C_2H_2NO$)

Vibrational Analysis — Indian J. Phys. 51B,
    399 (1977) — K.C. Medhi

o-Aminobenzaldehyde ($C_6H_4NH_2CHO$)

IR Spectrum — Ric. Sci. 24, 1687 (1954) —
    M. Scrocco and A. Liberti
IR Spectrum — J. Chem. Soc. 2051 (1959) —
    A.R. Katritzky and P. Simmons
IR Spectrum — J. Chem. Soc. 3670 (1959) —
    A.R. Katritzky and R.A. Jones
UV Absorption in Solutions — Can. J. Chem.
    38, 1837 (1960) — J.C. Dearden and
    W.F. Forbes
IR Spectrum — Indian J. Pure Appl. Phys.
    10, 50 (1972) — M.P. Srivastava, B.B.
    Lal and I.S. Singh

m-Aminobenzaldehyde ($C_6H_4NH_2CHO$)

IR Spectrum — Ric. Sci. <u>24</u>, 1687 (1954) —
    M. Scrocco and A. Liberti
IR Spectrum — J. Chem. Soc. 2051 (1959) —
    A.R. Katritzky and P. Simmons
IR Spectrum — J. Chem. Soc. 3670 (1959) —
    A.R. Katritzky and R.A. Jones
UV Absorption in Solutions — Can. J. Chem.
    <u>38</u>, 1837 (1960) — J.C. Dearden and
    W.F. Forbes
IR Spectrum — Indian J. Pure Appl. Phys.
    <u>10</u>, 50 (1972) — M.P. Srivastava, B.B.
    Lal and I.S. Singh

p-Aminobenzaldehyde ($C_6H_4NH_2CHO$)

IR Spectrum — Ric. Sci. <u>24</u>, 1687 (1954) —
    M. Scrocco and A. Liberti
IR Spectrum — J. Chem. Soc. 3670 (1959) —
    A.R. Katritzky and R.A. Jones
IR Spectrum — J. Chem. Sec. 2051 (1959) —
    A.R. Katritzky and P. Simmons
IR Spectrum — Indian J. Pure Appl. Phys.
    <u>10</u>, 50 (1972) — M.P. Srivastava, B.B.
    Lal and I.S. Singh

o-Aminobenzonitrile ($C_6H_4NH_2CN$)

UV Absorption in Solutions — J. Lumin. <u>10</u>,
    113 (1975) — Y.H. Lui and S.P. McGlynn
Fluorescence — J. Lumin. <u>10</u>, 113 (1975) —
    Y.H. Lui and S.P. McGlynn

m-Aminobenzonitrile ($C_6H_4NH_2CN$)

UV Absorption in Solutions — J. Lumin. <u>10</u>,
    113 (1975) — Y.H. Lui and S.P. McGlynn
Fluorescence — J. Lumin. <u>10</u>, 113 (1975) —
    Y.H. Lui and S.P. McGlynn

p-Aminobenzonitrile ($C_6H_4NH_2CN$)

Vibrational Analysis — Z. Elektrochem. <u>64</u>,
    1228 (1960) — S. Wackherlin and W.
    Luttke
Vibrational Analysis — Spectrochim. Acta
    <u>21</u>, 45 (1965) — H.W. Wilson and J.E.
    Bloor
UV Absorption in Splutions — J. Lumin. <u>10</u>,
    113 (1975) — Y.H. Lui and S.P. McGlynn
Fluorescence — J. Lumin. <u>10</u>, 113 (1975) —
    Y.H. Lui and S.P. McGlynn

m-Aminobenzotrifluoride ($C_6H_4NH_2CF_3$)

UV Absorption in Vapour — Indian J. Pure
    Appl. Phys. <u>7</u>, 831 (1969) — R. Ammi Amma,
    G.D. Baruah and K.P.R. Nair

p-Aminoazobenzene ($C_{12}H_{11}N_3$)

Raman (Resonance) — Spectrochim. Acta <u>32</u>,
    1179 (1976) — T. Uno, B-K. Kim, Y. Saito
    and K. Machida

3-Amino-2-Chlorotoluene ($C_6H_3CH_3ClNH_2$)

UV Absorption in Vapour — Indian J. Phys.
    <u>52B</u>, 125 (1978) — M.A. Shashidhar

3-Amino-4-Chlorotoluene ($C_6H_3CH_3ClNH_2$)

UV Absorption in Vapour — Indian J. Phys.
    <u>52B</u>, 125 (1978) — M.A. Shashidhar

6-Amino-3-Chlorotoluene ($C_6H_3CH_3ClNH_2$)

UV Absorption in Vapour — Indian J. Phys.
    <u>52B</u>, 125 (1978) — M.A. Shashidhar

1-Amino-3,5-Dimethylbenzene
($C_6H_3(CH_3)_2NH_2$)

See   Xylidine

2-Amino-4-Fluorotoluene ($C_6H_3CH_3NH_2F$)

Vibrational Analysis — Indian J. Pure Appl.
    Phys. <u>12</u>, 300 (1974) — S.J. Singh and
    S.M. Pandey

4-Amino-3-Fluorotoluene ($C_6H_3CH_3NH_2F$)

Vibrational Analysis — Indian J. Pure Appl.
    Phys. <u>12</u>, 300 (1974) — S.J. Singh and
    S.M. Pandey

5-Amino-2-Fluorotoluene ($C_6H_3CH_3NH_2F$)

Vibrational Analysis — Indian J. Pure Appl.
    Phys. <u>12</u>, 300 (1974) — S.J. Singh and
    S.M. Pandey

6-Amino-3-Fluorotoluene $(C_6H_3CH_3NH_2F)$

Vibrational Analysis — Indian J. Pure Appl.
Phys. 12, 300 (1974) — S.J. Singh and
S.M. Pandey

2-Aminomethylpyridine $(C_5H_3NCH_3NH_2)$

Vibrational Analysis — Curr. Sci. 34, 687
(1965) — K.V. Ramaiah and V.V. Chola-
pathi

2-Amino-3-Methylpyridine $(C_5H_3NCH_3NH_2)$

IR Spectrum — J. Indian Chem. Soc. 55, 699
(1978) — V.N. Verma

2-Amino-4-Methylpyridine $(C_5H_3NCH_3NH_2)$

IR Spectrum — J. Indian Chem. Soc. 55, 699
(1978) — V.N. Verma

2-Amino-4-Methylpyrimidine
$(C_4H_2N_2NH_2CH_3)$

IR Spectrum — Indian J. Pure Appl. Phys. 14,
842 (1976) — P.K. Goel, N.K. Sanyal and
S.L. Srivastava

2-Amino-3-Nitropyridine $(C_5H_3NNO_2NH_2)$

UV Absorption in Solution — Spectrochim.
Acta 23A, 89 (1967) — G. Favini, A.
Gambo and I.R. Bellobono

2-Amino-5-Nitropyridine $(C_5H_3NNO_2NH_2)$

UV Absorption in Solution — Spectrochim.
Acta 23A, 89 (1967) — G. Favini, A.
Gambo and I.R. Bellobono

3-Amino-2-Nitropyridine $(C_5H_3NNO_2NH_2)$

UV Absorption in Solution — Spectrochim.
Acta 23A, 89 (1967) — G. Favini, A.
Gambo and I.R. Bellobono

4-Amino-3-Nitropyridine $(C_5H_3NNO_2NH_2)$

UV Absorption in Solution — Spectrochim.
Acta 23A, 89 (1967) — G. Favini, A.
Gambo and I.R. Bellobono

Aminophenetole $(C_6H_4NH_2OC_2H_5)$

See   Ethoxyphenetole

o-Aminophenol $(C_6H_4OHNH_2)$

Raman — Monatsh. Chem. 76, 249 (1947) — E.
Herz and K.W.F. Kohlrausch
UV Absorption in Vapour — Indian J. Pure
Appl. Phys. 1, 230 (1963) — T.C. Sharma
and I.A. Rao
Vibrational Analysis — Appl. Spectrosc. 24,
445 (1970) — V.N. Verma and D.K. Rai

m-Aminophenol $(C_6H_4OHNH_2)$

Raman — Monatsh. Chem. 76, 249 (1947) —
E. Herz and K.W.F. Kohlrausch
Vibrational Analysis — Appl. Spectrosc. 24,
445 (1970) — V.N. Verma and D.K. Rai

p-Aminophenol $(C_6H_4OHNH_2)$

Raman — Monatsh. Chem. 72, 244 (1939) —
O. Paulsen
Vibrational Analysis — Spectrochim. Acta
21, 433 (1965) — R.J. Jakobsen
Vibrational Analysis — Appl. Spectrosc. 24,
445 (1970) — V.N. Verma and D.K. Rai

o-Aminopyridine $(C_5H_4NNH_2)$

UV Absorption in Solution — J. Amer. Chem.
Soc. 70, 3397 (1948) — E.A. Steck and
G.W. Ewing
UV Absorption in Solution — J. Amer. Chem.
Soc. 71, 340 (1949) — L.C. Anderson
and N.V. Seeger
Vibrational Analysis — J. Chem. Soc. 2911
(1952) — C.L. Angyal and R.L. Werner
Vibrational Analysis — J. Chem. Soc. 2939
(1952) — J.D.S. Goulden
Calculation — J. Chem. Phys. 22, 1077 (1954)
H.P. Stephenson
UV Absorption in Solution — J. Chem. Soc.
191 (1957) — A.R. Katritzky
UV Absorption in Solution — J. Chem. Soc.
4375 (1957) — J.M. Gardner and A.R.
Katritzky
Vibrational Analysis — J. Chem. Soc. 2202
(1958) — A.R. Katritzky and A.R. Hands
IR Spectrum (Hydrogen bonding) — J. Chem.
Soc. 3619 (1958) — S.F. Mason
Vibrational Analysis — J. Chem. Soc. 2198
(1958) — A.R. Katritzky and J.M. Gardner

IR Spectrum (Hydrogen bonding) – Spectrochim. Acta 16, 1176 (1960) – A.G.
Mortiz

UV Absorption in Different States – Indian
J. Phys. 35, 420 (1961) – T.N. Mishra

IR Spectrum – Appl. Spectrosc. 17, 90(1961)
R. Issac, F.F. Bentley, H. Sternglanz,
W.C. Coburn Jr., C.V. Stephenson and
W.S. Wilcox

IR Spectrum – J. Mol. Spectrosc. 7, 89
(1961) – K.V. Ramiah and P.G. Puranik

IR Spectrum (Hydrogen bonding) – Spectrochim. Acta 18, 671 (1962) – A.G.Mortiz

UV Absorption in Solution – Spectrochim.
Acta 23A, 89 (1967) – G. Favini, A.
Gambo and I.R. Bellobono

UV Absorption in Vapour – J. Chem. Phys.
65, 1486 (1968) – M. Lamotte and P.
Loustauneau

Vibrational Analysis – Indian J. Pure Appl.
Phys. 6, 104 (1968) – M.R. Padhye and
V.V. Bhujle

Phosphorescence – J. Phys. Chem. 72, 1982
(1968) – A. Weisstuch and A.C. Testa

UV Absorption in Vapour (High Resolution)
Mol. Phys. 18, 237 (1970) – J.M. Hollas,
G.H. Kirby and R.A. Wright

UV Absorption in Vapour – Indian J. Pure
Appl. Phys. 8, 820 (1970) – V.V. Bhujle
and M.R. Padhye

IR Spectrum – Indian J. Pure Appl. Phys.
8, 479 (1970) – M.R. Padhye and V.V.
Bhujle

Phosphorescence – J. Chem. Phys. 59, 596
(1973) – S. Hotchandani and A.C. Testa

o-Aminopyridine-ND$_2$ (C$_5$H$_4$NND$_2$)

UV Absorption in Vapour – Indian J. Pure
Appl. Phys. 8, 820 (1970) – V.V. Bhujle
and M.R. Padhye

IR Spectrum – Indian J. Pure Appl. Phys.
8, 479 (1970) – M.R. Padhye and V.V.
Bhujle

m-Aminopyridine (C$_5$H$_4$NH$_2$)

UV Absorption in Solution – J. Amer. Chem.
Soc. 70, 3397 (1948) – E.A. Steck and
G.W. Ewing

UV Absorption in Solution – J. Amer. Chem.
Soc. 71, 340 (1949) – L.C. Anderson
and N.V. Seeger

Vibrational Analysis – J. Chem. Soc. 2911
(1952) – C.L. Angyal and R.L. Werner

Vibrational Analysis – J. Chem. Soc. 2939
(1952) – J.D.S. Goulden

UV Absorption in Solution – J. Chem. Soc.
191 (1957) – A.R. Katrirzky

UV Absorption in Solution – J. Chem. Soc.
4375 (1957) – J.H. Gardner and A.R.
Katritzky

Vibrational Analysis – J. Chem. Soc. 2198
(1958) – A.R. Katritzky

Vibrational Analysis – J. Chem. Soc. 2202
(1958) – A.R. Katritzky and A.R. Hands

IR Spectrum (Hydrogen bonding) – J. Chem.
Soc. 3619 (1958) – S.F. Mason

IR Spectrum (Hydrogen bonding) – Spectrochim. Acta 16, 1176 (1960) – A.G. Mortiz

Vibrational Analysis – Appl. Spectrosc. 17,
90 (1961) – R. Issac, F.F. Bentley, H.
Sternglanz, W.C. Coburn Jr., C.V. Stephenson and W.S. Wilcox

UV Absorption in Different States – Indian
J. Phys. 35, 420 (1961) – T.N. Mishra

IR Spectrum – J. Mol. Spectrosc. 7, 89
(1961) – K.V. Ramiah and P.G. Puranik

IR Spectrum (Hydrogen bonding) – Spectrochim. Acta 18, 671 (1962) – A.G. Mortiz

UV Absorption in Solution – Spectrochim.
Acta 23A, 89 (1967) – G. Favini, A.
Gambo and I.R. Bellobono

Phosphorescence – J. Phys. Chem. 72, 1982
(1968) – A. Weisstuch and A.C. Testa

Vibrational Analysis – Indian J. Pure Appl.
Phys. 6, 104 (1968) – M.R. Padhye and
V.V. Bhujle

UV Absorption in Vapour – Indian J. Pure
Appl. Phys. 8, 820 (1970) – V.V. Bhujle
and M.R. Padhye

IR Spectrum – Indian J. Pure Appl. Phys. 8,
479 (1970) – M.R. Padhye and V.V. Bhujle

Phosphorescence – J. Chem. Phys. 59, 596
(1973) – S. Hotchandani and A.C. Testa

o-Aminothiophenol (C$_6$H$_4$SHNH$_2$)

Vibrational Analysis – Curr. Sci. 31, 79
(1962) – P.G. Puranik and V. Kumar

Aminotoluene (C$_6$H$_4$NH$_2$CH$_3$)

See   Methylaniline

Aniline (C$_6$H$_5$NH$_2$)

Raman – Monatsh. Chem. 74, 1 (1941) – K.W.F.
Kohlrausch and H. Wittek

Vibrational Analysis – Proc. Indian Acad.
Sci. A15, 390 (1942) – C.S. Venkateshwaran and N.S. Pandya

UV Absorption in Vapour – J. Chem. Phys. 13, 167 (1945) – N. Ginsburg and F.A. Matsen

Vibrational Analysis – Trans. Faraday Soc. 44, 767 (1948) – M.Sr.C. Flett

Vibrational Analysis – J. Chem. Phys. 20, 145 (1952) – N. Fuson, M.L. Josien, R.L. Powell and E. Utlerback

Vibrational Analysis – Spectrochim. Acta 9, 253 (1955) – D.H. Whiffen

Vibrational Analysis – Gazz. Chim. Ital. 87, 805 (1957) – S. Califano and R. Moceia

IR Spectrum – Spectrochim. Acta 13, 34 (1957) – L.J. Bellamy and R.L. Williams

IR Spectrum – Proc. Roy. Soc. A250, 22 (1959) P.J. Krueger and H.W. Thompson

Vibrational Analysis – C.R. Acad. Sci. (Paris) 248, 5604 (1959) – M.A. Kahane and J. Kahane-Paillous

UV Emission – J. Chim. Phys. 57, 1058 (1960)– J. Kahane-Paillous

Vibrational Analysis – Spectrochim. Acta 16, 428 (1960) – J.C. Evans

IR Spectrum (Hydrogen bonding) – Spectrochim. Acta 16, 559 (1960) – V.C. Farmer and R.H. Thomson

UV Absorption in Solutions – Can. J. Chem. 38, 1837 (1960) – J.C. Dearden and W.F. Forbes

Vibrational Analysis – Opt. Spektrosk. 11, 686 (1961) – V.S. Varshavshii

Vibrational Analysis – Ark. Fys. 21, 123 (1962) – F. Hjalmers

IR Spectrum – Indian J. Phys. 37, 139 (1963)– K.C. Medhi and G.S. Kastha

Vibrational Analysis – Spectrochim. Acta 19, 669 (1963) – G. Varsanyi, S. Holly and T. Farago

IR Spectrum – Spectrochim. Acta 22, 501 (1966) – W.R. McWhimie and K.C. Poller

UV Absorption in Vapour – J. Mol. Spectrosc. 20, 193 (1966) – J.C.D. Brand, D.R. Williams and T.J. Cook

UV Absorption in Vapour – J. Mol. Spectrosc. 20, 359 (1966) – J.C.D. Brand, D.R. Williams and T.J. Cook

UV Absorption in Vapour (High Resolution) – Mol. Phys. 16, 441 (1969) – J. Chritoffersen, J.M. Hollas and G.H. Kirby

Raman – J. Mol. Struct. 5, 477 (1970) – N.T. McDevitt and W.G. Fateley

Fluorescence – Ber. Bunsenges. Phys. Chem. 75, 450 (1971) – J.M. Blondeau and M. Stockburger

Fluorescence – J. Chem. Phys. 54, 2387 (1971) H. von Weyssenhoff and F. Kraus

Fluorescence – J. Mol. Spectrosc. 43, 87 (1972) – M. Quack and M. Stockburger

Fluorescence – Org. Mol. Photophys. 1, 57 (1973) – M. Stockburger

Calculation – J. Mol. Struct. 16, 365 (1973)– R.T.C. Brownlee, D.G. Cameron, R.D. Topson, A.R. Katritzky and A.J. Sparrow

Fluorescence – J. Chem. Phys. 61, 1782 (1974) – K.C. Smith, J.A. Schiavana and R.S. Freund

Fluorescence – J. Chem. Phys. 61, 1789 (1974) – K.C. Smith, J.A. Schiavana and R.S. Freund

Fluorescence – J. Chem. Phys. 61, 2160 (1974) – K.C. Smith, J.A. Schiavana and R.S. Freund

IR Spectrum – Spectrochim. Acta 30A, 835 (1974) – G. Brink and J.W. Bayles

Fluorescence – J. Chem. Phys. 61, 187 (1974)– W.R. Ware and A.M. Garcia

UV Absorption Spectrum – J. Mol. Spectrosc. 64, 139 (1977) – K. Fuke and S. Nagakura

Aniline–NHD (C_6H_5NHD)

Vibrational Analysis – Spectrochim. Acta 16, 428 (1960) – J.C. Evans

IR Spectrum – Spectrochim. Acta 30A, 835 (1974) – G. Brink and J.W. Bayles

Aniline–ND_2 ($C_6H_5ND_2$)

Vibrational Analysis – Spectrochim. Acta 16, 428 (1960) – J.C. Evans

Fluorescence – J. Mol. Spectrosc. 43, 87 (1972) – M. Quack and M. Stockburger

Fluorescence – Org. Mol. Photophys. 1, 57 (1973) – M. Stockburger

IR Spectrum – Spectrochim. Acta 30A, 835 (1974) – G. Brink and J.W. Bayles

Aniline–D_5 ($C_6D_5NH_2$)

Fluorescence – J. Mol. Spectrosc. 43, 87 (1972) – M. Quack and M. Stockburger

Aniline–D_7 ($C_6D_5ND_2$)

IR Spectrum – Curr. Sci. 39, 435 (1970) – J.N. Rai and R.C. Maheshwari

UV Absorption in Vapour – Indian J. Pure Appl. Phys. 9, 27 (1971) – J.N. Rai and R.C. Maheshwari

Fluorescence – J. Mol. Spectrosc. 43, 87 (1972) – M. Quack and M. Stockburger

Anisaldehyde ($C_6H_4CH_3CHO$)

See   Methoxybenzaldehyde

o—Anisidine ($C_6H_4OCH_3NH_2$)

Raman — Monatsh. Chem. <u>66</u>, 285 (1935) —
    K.W.F. Kohlrausch and G.P. Ypsilanti
Raman — Monatsh. Chem. <u>66</u>, 299 (1935) —
    A.W. Reitz and G.P. Ypsilanti
UV Absorption in Liquid — J. Amer. Chem. Soc.
    <u>76</u>, 5135 (1954) — H.E. Ungnade
IR Spectrum — Anal. Chem. <u>30</u>, 1598 (1958) —
    K.B. Whetsel, W.E. Robertson and M.W.
    Krell
UV Absorption in Liquid — Can. J. Chem. <u>37</u>,
    1294 (1959) — J.C. Dearden and W.F.
    Forbes
IR Spectrum — Anal. Chem. <u>32</u>, 217 (1960) —
    E.W. Stephen, C.B. Standey and H.B.
    Walter

m—Anisidine ($C_6H_4OCH_3NH_2$)

Raman — Monatsh. Chem. <u>66</u>, 285 (1935) —
    K.W.F. Kohlrausch and G.P. Ypsilanti
Raman — Monatsh. Chem. <u>66</u>, 299 (1935) —
    A.W. Reitz and G.P. Ypsilanti
UV Absorption in Liquid — J. Amer. Chem.Soc.
    <u>76</u>, 5135 (1954) — H.E. Ungnade
UV Absorption in Vapour — J. Sci. Ind. Res.
    (India) <u>168</u>, 230 (1957) — V.S. Rao and
    V.R. Rao
IR Spectrum — Anal. Chem. <u>30</u>, 1598 (1958) —
    K.B. Whetsel, W.E. Robertson and M.W.
    Krell
IR Spectrum — J. Chem. Soc. 2050 (1959) —
    A.R. Katritzky
UV Absorption in Solution — Can. J. Chem.
    <u>37</u>, 1294 (1959) — J.C. Dearden and
    W.F. Forbes
IR Spectrum — Anal. Chem. <u>32</u>, 217 (1960) —
    E.W. Stephen, C.B. Standey and H.B.
    Walter
IR Spectrum — Indian J. Phys. <u>37</u>, 139 (1963)
    K.C. Medhi and G.S. Kastha

p—Anisidine ($C_6H_4OCH_3NH_2$)

Raman — Monatsh. Chem. <u>66</u>, 285 (1935) —
    K.W.F. Kohlrausch and G.P. Ypsilanti
Raman — Monatsh. Chem. <u>66</u>, 299 (1935) —
    A.W. Reitz and G.P. Ypsilanti
IR Spectrum — J. Chem. Soc. 2310 (1952) —
    A. Burawoy and J.T. Chamberlain
UV Absorption in Liquid — J. Amer. Chem.
    Soc. <u>76</u>, 5135 (1954) — H.E. Ungnade
IR Spectrum — J. Chem. Soc. 4314 (1956) —
    A. Burawoy and J.T. Chamberlain

IR Spectrum — Anal. Chem. <u>30</u>, 1598 (1958) —
    K.B. Whetsel, W.E. Robertson and M.W.
    Krell
IR Spectrum — J. Chem. Soc. 2050 (1959) —
    A.R. Katritzky
UV Absorption in Liquid — Can. J. Chem. <u>37</u>,
    1294 (1959) — J.C. Dearden and W.F.
    Forbes
IR Spectrum — Anal. Chem. <u>32</u>, 217 (1960) —
    E.W. Stephen, C.B. Standey and H.B.
    Walter
UV Absorption in Liquid — Spectrochim. Acta
    <u>17</u>, 545 (1961) — E. Spinner

Anisole ($C_6H_5OCH_3$)

Raman — Monatsh. Chem. <u>65</u>, 6 (1934) — K.W.F.
    Kohlrausch, E. Herz and A. Pongratz
IR Spectrum — J. Phys. Radium <u>8</u>, 489 (1937)—
    J. Lecomte
IR Spectrum — J. Phys. Radium <u>10</u>, 143
    (1939) — P. Barchuvitz and M. Parodi
Raman — Monatsh. Chem. <u>72</u>, 244 (1939) — O.
    Paulsen
Raman — Monatsh. Chem. <u>76</u>, 231 (1946) —
    K.W.F. Kohlrausch
Raman — Monatsh. Chem. <u>76</u>, 112 (1946) — E.
    Herz, K.W.F. Kohlrausch and H.S. Albert
UV Absorption in Vapour — Curr. Sci. <u>19</u>, 48
    (1950) — K. Sreeramamurty
UV Absorption in Vapour — J. Amer. Chem. Soc.
    <u>72</u>, 1539 (1950) — W.W. Robertson, A.J.
    Seriff and F.A. Matsen
UV Absorption in Vapour — Indian J. Phys.
    <u>24</u>, 421 (1950) — K. Sreeramamurty
Vibrational Analysis — Indian J. Phys. <u>29</u>,
    503 (1955) — D.C. Biswas
Vibrational Analysis — Spectrochim. Acta
    <u>17</u>, 933 (1961) — C.V. Stephenson, W.C.
    Coburn Jr and W.S. Wilcox
IR Spectrum — J. Sci. Res. BHU (India) <u>13(1)</u>,
    44 (1962) — S. Prakash
Fluorescence — Nature <u>193</u>, 268 (1962) — S.
    Prakash
Vibrational Analysis — Spectrochim. Acta
    <u>18</u>, 39 (1962) — J.H.S. Green
Electronic Emission — Indian J. Phys. <u>37</u>,
    59 (1963) — S. Prakash and N.L. Singh
Fluorescence — Indian J. Phys. <u>37</u>, 59 (1963)—
    S. Prakash and N.L. Singh
IR Spectrum — Spectrochim. Acta <u>22</u>, 501
    (1965) — W.R. McWhinnie and R.C. Poller
Vibrational Analysis — Spectrochim. Acta
    <u>23A</u>, 1111 (1967) — M. Horak, E.R.
    Lippincott and R.K. Khanna
Vibrational Analysis — Spectrochim. Acta
    <u>25A</u>, 343 (1969) — N.L. Owen and R.E.
    Hester

Raman — J. Mol. Struct. <u>5</u>, 477 (1970) —
    N.T. McDevitt and W.G. Fateley
Emission Spectrum (Electron Impact) — Chem.
    Phys. <u>3</u>, 233 (1972) — T. Ogawa, M.
    Tsuji, M. Toyoda and N. Ishibashi

Anisole—D$_3$ (C$_6$H$_5$OCD$_3$)

Vibrational Analysis — J. Chem. Phys. <u>60</u>,
    1047 (1963) — M.T. Forel, C. Garrigou-
    Lagrange, J. Gemin and M.L. Josien

Anthracene (C$_{14}$H$_{10}$)

Fluorescence — Mol. Phys. <u>3</u>, 71 (1960) —
    B. Stevens and E. Hutton
Absorption Spectrum in Vapour — J. Mol.
    Spectrosc. <u>4</u>, 480 (1960) — L.E. Lyons
    and G.C. Morris
UV Absorption in Solution — Spectrochim. Acta
    <u>16</u>, 1060 (1960) — R.N. Jones and E.
    Spinner
IR Spectrum — Spectrochim. Acta <u>16</u>, 74
    (1960) — A.G. Moritz
Vibrational Analysis — J. Chem. Phys. <u>36</u>,
    903 (1962) — S. Califano
Vibrational Analysis — J. Chem. Phys. <u>41</u>,
    2575 (1964) — N. Abasbegovic, N. Vuholic
    and L. Colombo
Vibrational Analysis — Spectrochim. Acta <u>20</u>,
    547 (1964) — L. Colombo
IR Spectrum — Spectrochim. Acta <u>20</u>, 891
    (1964) — D.J. Evans and D.B. Scully
Absorption Spectrum — J. Mol. Spectrosc. <u>17</u>,
    24 (1965) — A.V. Bree
Fluorescence — J. Mol. Spectrosc. <u>17</u>, 24
    (1965) — A.V. Bree
Vibrational Analysis — Spectrochim. Acta <u>21</u>,
    217 (1965) — G.W. Chantry, A. Anderson,
    D.J. Browning and H.A. Gebbie
Absorption Spectrum — Can. J. Chem. <u>43</u>, 3253
    (1965) — J.P. Byrne and I.G. Ross
Fluorescence — J. Chem. Soc. Faraday II <u>69</u>,
    708 (1973) — O.L.J. Gizemann, F. Kaufman
    and G. Porter
Fluorescence — Proc. Roy. Soc. <u>A340</u>, 519
    (1974) — G.S. Beddard, G.R. Fleming,
    O.L.J. Gizemann and G. Porter

Anthracene—D$_{10}$ (C$_{14}$D$_{10}$)

Vibrational Analysis — Spectrochim. Acta <u>20</u>,
    547 (1964) — S. Colombo

9-Anthraldehyde (C$_{15}$H$_{10}$O)

IR Spectrum — Indian J. Pure Appl. Phys. <u>7</u>,
    649 (1969) — K. Singh, S. Nath Singh
    and D.K. Rai

Anthraquinone (C$_{14}$H$_8$O$_2$)

UV Absorption in Solution — Z. Naturforsch.
    <u>79</u>, 360 (1952) — H. Hartman and W. Lorenz
Emission Spectrum in Vapour — Indian J. Pure
    Appl. Phys. <u>5</u>, 197 (1967) — S. Nath Singh
    and R.S. Singh
UV Absorption in Vapour — Indian J. Pure
    Appl. Phys. <u>5</u>, 245 (1967) — S. Nath Singh
    and R.S. Singh

9,10-Anthraquinone (C$_{14}$H$_8$O$_2$)

UV Absorption in Vapour — J. Chem. Soc. 123
    (1923) — J.E. Purvis
UV Absorption in Solution — J. Chem. Soc. 159
    (1941) — R.A. Morton and W.T. Earlam
UV Absorption in Vapour — Indian Acad. Sci.
    <u>3A</u>, 159 (1941) — P.K. Seshan
UV Absorption in Solution — Z. Naturforsch.
    <u>79</u>, 360 (1952) — H. Hartman and W. Lorenz
IR Spectrum — Trans. Faraday Soc. <u>50</u>, 911
    (1954) — D. Hadzi and N. Sheppard
Emission Spectrum in Vapour — Indian J. Pure
    Appl. Phys. <u>5</u>, 342 (1967) — S. Nath Singh
    and R.S. Singh
UV Absorption in Vapour — Indian J. Pure Appl.
    Phys. <u>5</u>, 245 (1967) — S. Nath Singh and
    R.S. Singh
Emission Spectrum in Vapour — Indian J. Pure
    Appl. Phys. <u>6</u>, 91 (1968) — R.S. Singh
UV Absorption in Vapour — Indian J. Pure Appl.
    Phys. <u>8</u>, 641 (1970) — S. Nath Singh and
    R.S. Singh

Benzalchloride (C$_6$H$_5$CHCl$_2$)

UV Absorption in Vapour — J. Amer. Chem. Soc.
    <u>70</u>, 2842 (1948) — W.F. Harnner and F.A.
    Matsen

Benzaldehyde (C$_6$H$_5$CHO)

UV Absorption in Solution — J. Chem. Soc.
    2482 (1914) — P.E. Purvis
UV Absorption in Vapour — J. Chem. Soc. 2482
    (1914) — P.E. Purvis
Vibrational Analysis — Z. Phys. <u>79</u>, 455
    (1932) — R. Barr

UV Absorption in Vapour — J. Chim. Phys. 30, 528 (1933) — F. Almasy

Raman — Monatsh. Chem. 74, 253 (1943) — E. Herz, L. Kahovec and K.W.F. Kohlrausch

Vibrational Analysis — J. Chem. Phys. 19, 389 (1951) — S. Imanishi

UV Absorption in Vapour — J. Sci. Res. BHU (India) 2, 153 (1951) — S.N. Garg

UV Absorption in Vapour — Bull. Chem. Soc. Jap. 25, 150 (1952) — S. Imanishi, K. Semba, M. Ito and T. Anno

UV Absorption in Vapour — J. Chem. Phys. 20, 532 (1952) — S. Imanishi, M. Ito, K. Semba and T. Anno

UV Absorption in Vapour — J. Sci. Res. BHU (India) 4, 42, 62 (1953) — S.N. Garg

Vibrational Analysis — J. Sci. Res. BHU (India) 4, 68,83 (1953) — S.N. Garg

Emission Spectrum in Vapour — J. Chem. Phys. 22, 1384 (1954) — G.W. Robinson

Vibrational Analysis — Anal. Chem. 27, 2 (1955) — S. Pinchas

Raman — Indian J. Phys. 30, 530 (1956) — D.C. Biswas

Vibrational Analysis — Ann. Chem. 28, 1328 (1956) — D.F. Eggers, Jr

UV Emission in Vapour — J. Sci. Res. BHU (India) 9, 61 (1957) — S.N. Garg and I.S. Singh

UV Absorption in Solutions — Can. J. Chem. 38, 1877 (1960) — J.C. Dearden and W.F. Forbes

Raman — J. Chim. Phys. 58, 559 (1961) — C. Garrigou-Lagrange, N. Claverie, M.L. Josien and J.M. Lebas

Emission Spectrum — Z. Phys. Chem. 31, 350 (1962) — M. Stockburger

IR Spectrum — Nature 195, 595 (1962) — J.H.S. Green, W. Kynaston and H.A. Gebbie

IR Spectrum — Trans. Faraday Soc. 60, 5 (1964) — H.G. Silver and J.L. Wood

IR Spectrum — Spectrochim. Acta 22, 501 (1966) — W.R. McWhinnie and R.C. Poller

Raman — Spectrochim. Acta 23A, 462 (1967) — F.B. Brown

UV Absorption in Vapour — J. Chem. Phys. 49, 1745 (1968) — J.M. Hollas, E. Gregrek and L. Goodman

Vibrational Analysis — Indian J. Phys. 42, 610 (1968) — S. Chattopadhyaya and J. Jha

Raman — J. Mol. Struct. 5, 477 (1970) — N.T. McDevitt and W.G. Fateley

Vibrational Analysis — J. Mol. Spectrosc. 38, 336 (1971) — R. Zwarichm, J. Smolarek and L. Goodman

Phosphorescence — J. Mol. Spectrosc. 40, 71 (1971) — J. Olmsted III and M.A. El-Sayed

UV Absorption in Solution — Spectrochim. Acta 28A, 1969 (1972) — E. Vander Donckt and C. Vogels

Vibrational Analysis — J. Mol. Spectrosc. 43, 416 (1972) — J. Smolarek, R. Zwarich and L. Goodman

Calculation — J. Mol. Struct. 16, 365 (1973) — R.T.C. Brownlee, D.G. Cameron, R.D. Topsom, A.R. Katritzky and A.J. Sparrow

UV Absorption in Vapour — J. Amer. Chem. Soc. 95, 1717 (1973) — M. Berger, I.L. Goldblatt and C. Steel

UV Absorption in Vapour — Chem. Phys. 1, 385 (1973) — J.M. Hollas and S.N. Thakur

Emission Spectrum — Spectrochim. Acta 20, 1387 (1974) — Y. Kanda, H. Kasida and T. Matumura

UV Absorption in Vapour — J. Chem. Soc. Faraday Trans. II 71, 409 (1975) — J. Metcalf, R.G. Brown and D. Phillips

Excitation Spectrum in Vapour — J. Chem. Soc. Faraday Trans. II 71, 409 (1975) — J. Metcalf, R.G. Brown and D. Phillips

Vibrational Analysis — Spectrochim. Acta 32A, 1265 (1976) — J.H.S. Green and D.J. Harrison

UV Absorption in Vapour — Indian J. Phys. 51B, 184 (1977) — S.N. Thakur

Benzaldehyde-CDO (C_6H_5CDO)

Vibrational Analysis — Spectrochim. Acta 32A, 1265 (1976) — J.H.S. Green and D.J. Harrison

Benzaldehyde-D_1 (C_6H_4DCHO)

Vibrational Analysis — J. Mol. Spectrosc. 38, 336 (1971) — R. Zwarich, J. Smolarek and L. Goodman

Vibrational Analysis — J. Mol. Spectrosc. 39, 43 (1972) — J. Smolarek, R. Zwarich and L. Goodman

Benzaldehyde-D_4 (C_6HD_4CHO)

Vibrational Analysis — J. Mol. Spectrosc. 38, 336 (1971) — R. Zwarich, J. Smolarek and L. Goodman

Vibrational Analysis — J. Mol. Spectrosc. 39, 43 (1972) — J. Smolarek, R. Zwarich and L. Goodman

Benzaldehyde-D_6 (C_6D_5CDO)

Vibrational Analysis — J. Mol. Spectrosc. 38, 336 (1971) — R. Zwarich, J. Smolarek and L. Goodman

Vibrational Analysis — J. Mol. Spectrosc. 39, 43 (1972) — J. Smolarek, R. Zwarich and L. Goodman

Benzene   (C_6H_6)

Vibrational Analysis — Phys. Rev. 45, 706 (1934) — E.B. Wilson

Vibrational Analysis — J. Chem. Soc. 925 (1936) — W.G. Angus, C.K. Ingold and A.H. Leckie

Vibrational Analysis — J. Chem. Soc. 912 (1936) — C.K. Ingold

UV Absorption in Vapour — J. Chem. Phys. 5, 669 (1937) — A.L. Sklar

UV Absorption in Vapour — J. Chem. Phys. 7, 207 (1939) — H. Sponer, G. Nordheim, A.L. Sklar and E. Teller

UV Absorption in Vapour — J. Chem. Phys. 8, 905 (1940) — H. Sponer

UV Absorption in Vapour — Rev. Mod. Phys. 13, 115 (1941) — H. Sponer and E. Teller

UV Absorption in Vapour — Rev. Mod. Phys. 14, 224 (1942) — H. Sponer

UV Absorption in Vapour — Rev. Mod. Phys. 14, 232 (1942) — A.L. Sklar

Vibrational Analysis — J. Chem. Soc. 222 (1946) — C.K. Ingold

Vibrational Analysis — J. Chem. Phys. 14, 282 (1946) — F.A. Miller and B. Crawford

Raman — J. Chem. Soc. 391 (1949) — I. Ichishima

Raman — J. Chem. Phys. 18, 1119 (1950) — A. Fruchling

Structural Data — Acta Crystallogr. 3, 46 (1950) — P.W. Allen and L.E. Sutton

UV Absorption in Vapour — Helvchim. Acta 34, 462 (1951) — F. Almasy and H.Laemmel

Raman — Ann. Phys. 6, 401 (1951) — A. Fruhling

Calculation — Phil. Trans. Roy. Soc. A248, 131 (1955) — D.H. Whiffen

UV Absorption in Vapour — J. Chem. Phys. 22, 234 (1955) — H. Sponer

Raman (Solid) — Indian J. Phys. 30, 313 (1956) — G.S. Kastha

Vibrational Analysis — Mat. Fys. Skr. Kong. Densk. Vid. Selk. 1, 1 (1956) — S. Brodersen and A. Langseth

IR Spectrum — J. Phys. Chem. 61, 730 (1957) — S.H. Hestings ans D.E. Nichalson

IR Spectrum — Spectrochim. Acta 13, 180 (1958) — A. Danti and R.C. Lord

Calculation — J. Mol. Spectrosc. 5, 238 (1960) — A.C. Albrecht

Phosphorescence — Spectrochim. Acta 16, 1135 (1960) — H. Sponer, Y. Kanda and L.A. Blackwell

UV Absorption in Vapour — Indian J. Phys. 34, 581 (1960) — S.C. Sirkar and J.K. Roy

UV Absorption in Solution — Spectrochim. Acta 17, 424 (1961) — Y. Kanda, Y. Gondo and R. Shimada

Calculations — J. Mol. Spectrosc. 6, 497 (1961) — E. Elementi

UV Absorption in Vapour — J. Mol. Spectrosc. 7, 304 (1961) — S. Leach, R. Lopez-Delgado and F. Delmas

Phosphorescence — Spectrochim. Acta 17, 7 (1961) — Y. Kanda and R. Shimada

Phosphorescence — Spectrochim. Acta 17, 1298 (1961) — J.D. Spangler and H. Sponer

Fluorescence — J. Chem. Phys. 35, 1389 (1961) — A.B.F. Duncan and J.W. Donovan

IR Spectrum — Spectrochim. Acta 18, 1287 (1962) — N.S. Baylis and N.W. Cant

UV Absorption (Crystal) — J. Mol. Spectrosc. 9, 288 (1962) — J.R. Platt

Fluorescence — J. Chem. Phys. 37, 583 (1962) — H. Ishikawa and W.A. Noyes, Jr

Calculation — Spectrochim. Acta 19, 601 (1963) — J.R. Scherer

Raman (Crystal) — Indian J. Phys. 38, 181 (1964) — S.C. Sirkar, D.K. Mukherjee and P.K. Bishnui

Vibrational Analysis — Spectrochim. Acta 20, 345 (1964) — J.R. Scherer

Raman — Indian J. Phys. 38, 174 (1964) — S.C. Sirkar, D.K. Mukherjee and P.K. Bishnui

Raman — Opt. Spektrosk. 16, 128 (1964) — M.M. Sushchinski and Z.M. Muldakhmetov

Calculation — Spectrochim. Acta 21, 321 (1965) — J.R. Scherer

Fluorescence — J. Chem. Phys. 42, 2942 (1965) — G.B. Kistiakowsky and C.S. Parmenter

UV Absorption Spectrum — J. Mol. Spectrosc. 18, 158 (1965) — Y. Diamont, R.M. Hexter and O. Schnepp

Fluorescence — Proc. Roy. Soc. A283, 83 (1965) — J.B. Birks, C.L. Braga and M.D. Lumb

UV Absorption in Vapour — J. Mol. Spectrosc. 15, 394 (1965) — G.W. King and E.H. Pinnington

Excitation Spectrum (Two Photon) — J. Chem. Phys. 46, 2714 (1966) — B. Honig, J. Jortner and A. Szoke

UV Absorption Spectrum — J. Mol. Spectrosc. 19, 456 (1966) — E.W. Hollier and C.E. Blount

16

UV Absorption in Vapour – J. Mol. Spectrosc. 20, 96 (1966) – A. Grabowska

Vibrational Analysis – J. Chem. Phys. 44, 2016 (1966) – I. Harada and T. Shimanouchi

Vibrational Analysis – Spectrochim. Acta 22, 1029 (1966) – M. Ito and T. Shigeoka

UV Absorption in Vapour – Philos. Trans. R. Soc. Lond. Ser A 259, 499 (1966) – J.H. Colloman, T.M. Dunn and I.M. Mills

Vibrational Analysis – Spectrochim. Acta 23A, 1489 (1967) – J.R. Scherer

UV Absorption in Vapour – C.R. Acad. Sci. B265, 641 (1967) – A. Quemerais, M. Merlais and S. Robin

Vibrational Analysis (Crystal) – J. Chem. Phys. 46, 2708 (1967) – I. Harada and T. Shimanouchi

Calculation – J. Mol. Spectrosc. 22, 296 (1967) – W.D. Jones

Fluorescence – J. Chem. Phys. 44, 2100 (1968) – W.A. Noyes, Jr., W.A. Mulac and D.A. Harter

Fluorescence – J. Chem. Phys. 48, 4748 (1968) – E.M. Anderson and G.B. Kistiakowski

Calculation – J. Chem. Phys. 49, 2261 (1968) – R.J. Bunker, J.L. Whitten and J.D. Petke

UV Absorption in Solutions – J. Mol. Spectrosc. 25, 273 (1968) – M. Koyanagi

Fluorescence – J. Chem. Phys. 50, 1631 (1969) – C.S. Parmenter and A.H. White

Fluorescence – J. Chem. Phys. 51, 3130, 3615 (1969) – G.M. Breuer and E.K.C. Lee

Vibrational Analysis – C.R. Acad. Sci. (Paris) – B268, 1366 (1969) – B. Pasquier, C. Sourisseau and M.L. Josien

Fluorescence – J. Chem. Phys. 51, 1551 (1969) – C.S. Parmenter and H.M. Poland

IR Spectrum (Intensity) – J. Mol. Spectrosc. 32, 265 (1969) – D. Steele and W. Wheatley

Raman – J. Mol. Struct. 3, 242 (1969) – H.W. Schrotter and J. Bofilias

Fluorescence – J. Chem. Phys. 52, 107 (1970) – M. Nishikawa and P.K. Ludwig

Fluorescence – Chem. Phys. Lett. 6, 345 (1970) – W. Gelbart, K.G. Spears, K.F. Freed, J. Jortner and S.A. Rice

UV Absorption in Inert Solid – J. Mol. Spectrosc. 35, 61 (1970) – G. Smith, S. Henry and C.E. Blount

IR Spectrum – J. Phys. B– Atom. Mol. Phys. 31, 8510 (1970) – U. Kumar

Fluorescence – Chem. Phys. Lett. 6, 352 (1970) – C.S. Burton and H.E. Hunziker

Fluorescence – Chem. Phys. Lett. 6, 339 (1970) – C.S. Parmenter and M.W. Schuyler

Fluorescence – J. Chem. Phys. 52, 3502 (1970) – C.S. Burton and H.E. Hunziker

Fluorescence – J. Chem. Phys. 52, 107 (1970) – M. Nishikawa and P.K. Ludwig

Fluorescence – Chem. Phys. Lett. 6, 339 (1970) – C.S. Parmenter and M.W. Schuyler

Fluorescence – J. Chem. Phys. 52, 5482 (1970) – B.K. Sellinger and W.R. Ware

UV Absorption Spectrum – J. Chem. Phys. 52, 88 (1970) – B. Katz, M. Brith, B. Sharf and J. Jortner

Fluorescence – J. Chem. Phys. 53, 268 (1970) – H.F. Kemper and M. Stockburger

Fluorescence – J. Chem. Phys. 53, 3160 (1970) – B.K. Sellinger and W.R. Ware

Fluorescence – J. Chem. Phys. 53, 5366 (1970) – C.S. Parmenter and M.W. Schuyler

Raman – J. Mol. Struct. 5, 477 (1970) – N.T. McDevitt and W.G. Fateley

Fluorescence – Ber. Bunsenges. Phys. Chem. 75, 450 (1971) – J.M. Blondeau and M. Stockburger

Fluorescence – J. Chem. Phys. 54, 4169 (1971) – W.H. Smith

Fluorescence – J. Chem. Phys. 55, 5561 (1971) – S.A. Rice and K.G. Spears

Fluorescence – J. Chem. Phys. 75, 1572 (1971) – G.H. Atkinson, C.S. Parmenter and M.W. Schuyler

UV Absorption in Vapour – J. Phys. Chem. 75, 1564 (1971) – G.H. Atkinson and C.S. Parmenter

UV Absorption in Vapour – J. Chem. Phys. 54, 3924 (1971) – B. Katz, M. Brith, B. Sharf and J. Jortner

Fluorescence – J. Chem. Phys. 56, 2309 (1972) – D.F. Heller, K.F. Freed and W.M. Gelbart

Fluorescence – Chem. Phys. Lett. 14, 404 (1972) – G.M. Breuer and E.K.C. Lee

Raman – J. Chem. Phys. 57, 99 (1972) – H. Bonadeo, M.P. Marzocchi, E. Castellucci and S. Califano

Fluorescence – Israel J. Chem. 10, 72 (1972) – M. Luria

Fluorescence – Adv. Chem. Phys. 22, 365 (1972) – C.S. Parmenter

UV Absorption in Solid – Chem. Phys. Lett. 17, 588 (1972) – E. Pantos and T.D.S. Hamilton

IR Spectrum (HCl Matrix) – Spectrochim. Acta 28A, 15 (1972) – K. Szczepaniak and W.B. Person

Fluorescence - Org. Mol. Photophys. $\underline{1}$, 57
   (1973) - M. Stockburger
Emission Spectrum (Electron Impact) -
   Bull. Chem. Soc. Jap. $\underline{46}$, 2637 (1973) -
   T. Ogawa, M. Tsuji, M. Toyoda and N.
   Ishibashi
Excitation Spectrum - Chem. Phys. Lett.
   $\underline{24}$, 1 (1973) - R.M. Hochstrasser and
   J.E. Wessel
UV Absorption Spectrum (E_{2g}) - J. Mol.
   Spectrosc. $\underline{45}$, 271 (1973)- G.C. Morris
   and J.G. Angus
Calculation - J. Mol. Spectrosc. $\underline{46}$, 207
   (1973) - R.J. Hayward, B.R. Henry and
   W. Siebrand
Calculation - J. Mol. Spectrosc. $\underline{48}$, 446
   (1973) - V.J. Eaton and D. Steele
Calculation - J. Mol. Struct. $\underline{16}$, 365
   (1973) - R.T.C. Brownlee, D.G. Cameron,
   R.D. Topsom, A.R. Katritzky and A.J.
   Sparrow
Fluorescence - J. Chem. Phys. $\underline{61}$, 4747
   (1974) - K.C. Smith, J.A. Schiavone
   and R.S. Freund
Fluorescence - Chem. Phys. $\underline{6}$, 331 (1974) -
   M.G. Prais, D.F. Heller and K.F.
   Freed
Fluorescence (Electron Impact) - J. Chem.
   Phys. $\underline{61}$, 1789 (1974) - K.C. Smith,
   J.A. Schiavone and R.S. Freund
Fluorescence - J. Phys. Chem. $\underline{78}$, 1904
   (1974) - M. Lucia, M. Ofran and G.
   Stein
Raman (Solid) - J. Chem. Phys. $\underline{61}$, 1380
   (1974) - W.D. Ellenson and M. Nicol
Fluorescence - J. Chem. Phys. $\underline{61}$, 1782
   (1974) - K.C. Smith, J.A. Schiavone
   and R.S. Freund
Fluorescence - J. Chem. Phys. $\underline{61}$, 1789,
   2160 (1974) - K.C. Smith, J.A. Schia-
   vone and R.S. Freund
Fluorescence - Chem. Phys. Lett. $\underline{28}$, 320
   (1974) - C.J.M. Beenakker and L.J.
   Oosterhoff
Excitation Spectrum (Two Photon) - J.
   Chem. Phys. $\underline{60}$, 317 (1974) - R.M.
   Hochstrasser, J.E. Wessel and H.N.
   Sung
Fluorescence - J. Amer. Chem. Soc. $\underline{97}$,
   1993 (1975) - A.E.W. Knight, C.S.
   Parmenter and M.W. Schuyler
Excitation Spectrum - Chem. Phys. Lett.
   $\underline{32}$, 210 (1975) - L. Wunsch, H.J.
   Neusser and E.W. Schlag
Excitation Spectrum - Chem. Phys. Lett.
   $\underline{31}$, 433 (1975) - L. Wunsch, H.J.
   Neusser and E.W. Schlag
Excitation Spectrum - Chem. Phys. Lett.
   $\underline{34}$, 109 (1975) - F. Metz

Fluorescence - J. Chem. Phys. $\underline{62}$, 573
   (1975) - R. Azria and G.J. Schulz
Calculation - Mol. Phys. $\underline{29}$, 673 (1975) -
   I.N. Douglas, R. Grinter and A.J.
   Thomson
Fluorescence - J. Amer. Chem. Soc. $\underline{97}$,
   1993 (1975) - C.S. Parmenter, A.F.W.
   Knight and M.W. Schuyler
IR Spectrum (Inert Gas Matrices) -
   Spectrochim. Acta $\underline{32A}$, 145 (1976) -
   M. Spoleti, S.N. Cesaro and V. Crosso
IR Spectrum - Spectrochim. Acta $\underline{32A}$, 1545
   (1976) - S.J. Daunt and S.F. Shurvell
Phosphorescence - J. Mol. Spectrosc. $\underline{60}$,
   71 (1976) - L. Hellner and C. Vermeil
Fluorescence - J. Mol. Spectrosc. $\underline{60}$,
   71 (1976) - L. Hellner and C. Vermeil
Fluorescence - J. Mol. Spectrosc. $\underline{62}$, 313
   (1976) - L. Hellner and C. Vermeil
Fluorescence - Chem. Phys. Lett. $\underline{43}$, 228
   (1976) - M.D. Swords and D. Phillips
Phosphorescence - J. Mol. Spectrosc. $\underline{62}$,
   313 (1976) - L. Hellner and C. Vermeil
Calculation - J. Mol. Struct. $\underline{31}$, 187
   (1976) - J.T. Gleghorn, S. Hadjipavlou
   and F.W. McConkey
Calculation - J. Mol. Struct. $\underline{32}$, 93
   (1976) - F. Torok, A. Hegedus, K. Kosa
   and P. Pulay
IR Spectrum (Matrix) - Spectrochim. Acta
   $\underline{34A}$, 117 (1978) - K.G. Brown and W.B.
   Person
Calculation - Spectrochim. Acta $\underline{34A}$, 1019
   (1978) - P.C. Painter and J.L. Koenig

Benzene-D_1 (C_6H_5D)

Vibrational Analysis - J. Chem. Soc. 299
   (1946) - C.R. Bailey, R.R. Gordon,
   J.B. Hale, N. Herzfeld, C.K. Ingold
   and H.G. Poole
Vibrational Analysis - Opt. Spektrosk.
   $\underline{26}$, 1034 (1969) - A.V. Korshunov
   and L.I. Namizerova

Benzene-p-D_2 ($C_6H_4D_2$)

Vibrational Analysis - J. Chem. Soc. 272
   (1946) - N. Herzfeld, J.W. Hobden,
   C.K. Ingold and H.G. Poole
Vibrational Analysis - J. Chem. Soc. 288
   (1946) - C.R. Bailey, S.C. Carson,
   R.R. Gordon and C.K. Ingold
Normal Coordinate Analysis - J. Mol.
   Spectrosc. $\underline{5}$, 236 (1960) - A.C. Albr-
   echt
Vibrational Analysis - Opt. Spektrosk.
   $\underline{26}$, 1034 (1969) - A.V. Korshunov and
   L.I. Namizerova

Benzene—1,3,5—D_3 ($C_6H_3D_3$)

IR Spectrum — J. Chem. Soc. 255 (1946) — C.R. Bailey, J.B. Hale, N. Herzfeld, C.K. Ingold, A.H. Leckie and H.G. Poole

IR Spectrum — Trans. Faraday Soc. 46, 103 (1950) — A.R.H. Cole and H.W. Thompson

IR Spectrum — Spectrochim. Acta 13, 180 (1958) — A. Danti and R.C. Lord

IR Spectrum — Bull. Chem. Soc. Jap. 33, 1024 (1960) — S. Saeki

Normal Coordinate Analysis — J. Mol. Spectrosc. 5, 236 (1960) — A.C. Albrecht

Vibrational Analysis — Spectrochim. Acta 20, 345 (1964) — J.R. Scherer

Vibrational Analysis — Spectrochim. Acta 23A, 1489 (1967) — J.R. Scherer

Calculation — J. Mol. Spectrosc. 48, 466 (1973) — V.J. Eaton and D. Steele

Benzene—1,2,4,5—D_4 ($C_6H_2D_4$)

IR Spectrum — J. Chem. Soc. 272 (1946) — N. Herzfeld, J.W. Hobden, C.K. Ingold and H.G. Poole

IR Spectrum — J. Chem. Soc. 288 (1946) — C.R. Bailey, S.C. Carson, R.R. Gordon and C.K. Ingold

Vibrational Analysis — Spectrochim. Acta 19, 1955 (1963) — D.A. Long and D. Steele

Benzene—2,3,5,6—D_4 ($C_6H_2D_4$)

Normal Coordinate Analysis — J. Mol. Spectrosc. 5, 236 (1960) — A.C. Albrecht

Benzene—D_6 (C_6D_6)

IR Spectrum — J. Chem. Soc. 966 (1946) — W.G. Angus, C.R. Bailey, J.B. Hale, C.K. Ingold, A.H. Leckie, C.G. Raisin J.W. Thompson and C.L. Wilson

IR Spectrum — Spectrochim. Acta 13, 180 (1958) — A. Danti and R.C. Lord

Normal Coordinate Analysis — J. Mol. Spectrosc. 5, 236 (1960) — A.C. Albrecht

Calculation — Spectrochim. Acta 18, 433 (1962) — D.A. Dows and A.L. Platt

Vibrational Analysis — Spectrochim. Acta 20, 345 (1964) — J.R. Scherer

Vibrational Analysis — Opt. Spektrosk. 16, 234 (1964) — M.M. Suchchinski and E.M. Muebaklmetov

UV Absorption in Vapour — J. Mol. Spectrosc. 15, 394 (1965) — G.W. King and E.H. Pinnington

Vibrational Analysis — Spectrochim. Acta 23A, 1489 (1967) — J.R. Scherer

Raman — J. Mol. Struct. 3, 242 (1969) — H.W. Schrotter and J. Bofilias

Vibrational Analysis — C.R. Acad. Sci. (Paris) 268B, 1366 (1969) — B. Pasquier, C. Sourisseau and M.L. Josien

Vibrational Analysis — Spectrochim. Acta 26A, 1097 (1970) — W.D. Mross and G. Zundel

Raman — J. Mol. Struct. 5, 477 (1970) — N.T. McDevitt and W.G. Fateley

Fluorescence — J. Chem. Phys. 52, 5366 (1970) — C.S. Parmenter and M.W. Schuyler

Fluorescence — Chem. Phys. Lett. 6, 352 (1970) — C.S. Burton and H.E. Hunziker

Fluorescence — J. Chem. Phys. 53, 3160 (1970) — B.K. Sellinger and W.R. Ware

Fluorescence — J. Chem. Phys. 56, 2309 (1972) — D.F. Heller, K.F. Freed and W.M. Gilbert

Fluorescence — J. Chem. Phys. 56, 2291 (1972) — A.S. Abramson, K.G. Spears and S.A. Rice

Calculation — J. Mol. Spectrosc. 48, 446 (1973) — V.J. Eaton and D. Steele

Fluorescence — Chem. Phys. Lett. 31, 433 (1975) — L. Wunsch, H.J. Neusser and E.W. Schlag

IR Spectrum (Inert Gas Matrices) — Spectrochim. Acta 32A, 145 (1976) — M. Spoliti, S. Cesaro and V. Grosso

IR Spectrum — Spectrochim. Acta 32A, 1545 (1976) — S.J. Daunt and H.F. Shurvell

Fluorescence — J. Mol. Spectrosc. 60, 71 (1976) — L. Hellner and C. Vermeil

Phosphorescence — J. Mol. Spectrosc. 60, 71 (1976) — L. Hellner and C. Vermeil

Calculation — J. Mol. Struct. 32, 93 (1976) — F. Torok, A. Hegedus, K. Kosa and P. Puley

Benzenethiol (C_6H_5SH)

IR Spectrum — Indian J. Phys. 5, 219 (1930)— S. Venkateshwaran

Vibrational Analysis — J. Amer. Chem. Soc. 78, 5463 (1956) — D.W. Scott, J.P. McCullough, W.N. Hubbard, J.F. Messerley, I.A. Hossenlopp, F.R. Frow and G. Waddington

IR Spectrum — J. Mol. Struct. 22, 29 (1974)— N.W. Larsen and F.M. Nicolaisen

1,4-Benzodiazene ($C_8H_5N_3$)

See   Quinoxaline

Benzonitrile (C_6H_5CN)

Raman — Monatsh. Chem. 63, 427 (1933) —
    K.W.F. Kohlrausch and A. Pongratz
UV Absorption in Vapour — Bull. Chem. Soc.
    Jap. 11, 346 (1936) — K. Masaki
IR Spectrum — C.R. Acad. Sci. (Paris) —
    200B, 30 (1939) — R. Barchewitz and
    M. Parodi
UV Absorption in Vapour — J. Chem. Phys.
    16, 480 (1948) — R.C. Hirt and J.P.
    Howe
Fluorescence — J. Chem. Phys. 18, 1403
    (1950) — A.M. Bass
Emission Spectrum in Vapour — Curr. Sci.
    25, 150 (1956) — R.K. Asundi and B.D.
    Joshi
Vibrational Analysis — Z. Elektrochem. 62,
    544 (1958) — J. Behringer
Vibrational Analysis — Z. Elektrochem. 64,
    560 (1960) — B. Bak and J.T. Nielsen
Emission Spectrum — J. Chim. Phys. 57, 581,
    1048 (1960) — J. Kahane-Paillous
Vibrational Analysis — Z. Elektrochem. 64,
    1228 (1960) — S. Weckherlin and W.
    Luttke
Vibrational Analysis — Spectrochim. Acta
    17, 607 (1961) — J.H.S. Green
Phosphorescence — Spectrochim. Acta 18, 1201
    (1962) — K. Takei and Y. Kanda
Vibrational Analysis — J. Chim. Phys. 59,
    1072 (1962) — J.M. Lebas
Vibrational Analysis — Spectrochim. Acta
    21, 45 (1965) — H.W. Wilson and J.E.
    Bloor
Vibrational Analysis — Spectrochim. Acta
    21, 127 (1965) — R.J. Jackobsen
IR Spectrum — Spectrochim. Acta 22, 501
    (1966) — W.R. McWhinnie and R.C. Poller
Phosphorescence — C.R. Acad. Sci (Paris) —
    269B, 431 (1969) — E.Faure, F. Valadier
    and J. Janin
UV Absorption in Vapour (High Resolution) —
    J. Mol. Spectrosc. 36, 328 (1970) —
    J.C.D. Brand and P.D. Knight
Raman — J. Mol. Struct. 5, 477 (1970) — N.T.
    McDevitt and W.G. Fateley
Calculation — Opt. Spektrosk. 31, 341
    (1971) — K.M. Danchinov, A.N. Rodinov,
    E.A. Gastilovich and D.N. Shigorin
Phosphorescence — J. Mol. Spectrosc. 41,
    249 (1972) — G.L. LeBel and J.D. Laposa
Fluorescence — J. Mol. Spectrosc. 41, 249
    (1972) — G.L. LeBel and J.D. Laposa

Calculation — J. Mol. Struct. 16, 365
    (1973) — R.T.C. Brownlee, D.G. Cameron,
    R.D. Topsom, A.R. Katritzky and A.J.
    Sparrow
Vibrational Analysis — Spectrochim. Acta
    32A, 1279 (1976) — J.H.S. Green and
    D.J. Harrison

Benzonitrile-o-D_1 (C_6H_4CND)

Vibrational Analysis — Z. Elektrochem.
    64, 560 (1960) — B. Bak and J.T.
    Nielsen
Phosphorescence — J. Mol. Spectrosc. 41,
    249 (1972) — G.L. LeBel and J.D.
    Laposa
Fluorescence — J. Mol. Spectrosc. 41,
    249 (1972) — G.L. LeBel and J.D.
    Laposa

Benzonitrile-m-D_1 (C_6H_4CND)

Vibrational Analysis — Z. Elektrochem.
    64, 560 (1960) — B. Bak and J.T.
    Nielsen
Fluorescence — J. Mol. Spectrosc. 41, 249
    (1972) — G.L. LeBel and J.D. Laposa
Phosphorescence — J. Mol. Spectrosc. 41,
    249 (1972) — G.L. LeBel and J.D.
    Laposa

Benzonitrile-p-D_1 (C_6H_4CND)

Vibrational Analysis — Z. Elektrochem.
    64, 560 (1960) — B. Bak and J.T.
    Nielsen
Fluorescence — J. Mol. Spectrosc. 41, 249
    (1972) — G.L. LeBel and J.D. Laposa
Phosphorescence — J. Mol. Spectrosc. 41,
    249 (1972) — G.L. LeBel and J.D.
    Laposa

Benzonitrile-D_5 (C_6D_5CN)

Vibrational Analysis — Spectrochim. Acta
    21, 127 (1965) — R.J. Jackobsen
Fluorescence — J. Mol. Spectrosc. 41, 249
    (1972) — G.L. LeBel and J.D. Laposa
Phosphorescence — J. Mol. Spectrosc. 41,
    249 (1972) — G.L. LeBel and J.D.
    Laposa

Benzopyrimidine ($C_8H_6N_2$)

See   Quinaxoline

Benzoquinoe ($C_6H_4O_2$)

Absorption Spectrum in Solution — J. Chem. Soc. 44, 490 (1945) — E.A. Braude

Emission Spectrum — Spectrochim. Acta 4, 85 (1950) — H. Schuler

UV Absorption in Solution — J. Chem. Phys. 20, 700 (1952) — H. McConnell

UV Absorption in Vapour — Nature 176, 1223 (1955) — R.K. Asundi and R.S. Singh

Emission Spectrum — J. Amer. Chem. Soc. 78, 2363 (1956) — J.W. Sidmann

UV Absorption in Vapour — J. Chem. Phys. 26, 967 (1957) — T. Anno and A. Sado

UV Absorption in Vapour — J. Sci. Res. BHU (India) 8, 261 (1957) — R.S. Singh

UV Absorption in Vapour — Indian J. Phys. 33, 376 (1959) — R.S. Singh

UV Absorption in Vapour — J. Chem. Phys. 32, 1602 (1960) — T. Anno and A. Sado

Absorption Spectrum in Solution — Bull. Chem. Soc. Jap. 35, 295 (1962) — A. Kuboyama

Visible Absorption Spectrum — Bull. Chem. Soc. Jap. 35, 1520 (1962) — A. Sado

Vibrational Analysis — Trans. Faraday Soc. 59, 1248 (1963) — M. Davies and F.E. Prichard

Visible Emission Spectrum in Vapour — Indian J. Pure Appl. Phys. 2, 358 (1964) — M.G. Jayswal and R.S. Singh

UV Absorption in Vapour — Spectrochim. Acta 20, 1563 (1964) — J.M. Hollas

UV Absorption in Vapour — J. Chem. Phys. 43, 760 (1965) — J.M. Hollas and L. Goodman

Calculation — J. Mol. Spectrosc. 17, 58 (1965) — P.E. Stevenson

Emission in Vapour — J. Mol. Spectrosc. 17, 6 (1965) — M.G. Jayswal and R.S. Singh

Emission Spectrum in Vapour — Spectrochim. Acta 21, 1597 (1965) — M.G. Jayswal and R.S. Singh

Visible Spectrum (intensity) — J. Mol. Spectrosc. 20, 312 (1966) — I.G. Ross, J.M. Hollas and K.K. Innes

Absorption Spectrum — J. Mol. Spectrosc. 23, 191 (1967) — P.E. Stevenson

Emission Spectrum in Vapour — Indian J. Pure Appl. Phys. 6, 91 (1968) — R.S. Singh

Emission Spectrum in Vapour — Spectrochim. Acta 25A, 39 (1968) — G. Breigleb, W. Herre and D. Wolf

UV Absorption in Vapour (High Resolution) Mol. Phys. 17, 655 (1969) — J. Christ-offersen and J.M. Hollas

Emission Spectrum — J. Mol. Spectrosc. 34, 450 (1970) — M. Koyanagi, Y. Kogo and Y. Kanda

Calculation — J. Mol. Struct. 14, 75 (1972) — O.F. Bizri and F.E. Prichard

Fluorescence — J. Mol. Spectrosc. 50, 14 (1974) — T.M. Dunn and A.H. Francis

Vibrational Analysis — J. Mol. Spectrosc. 50, 1 (1974) — T.M. Dunn and A.H. Francis

Calculation — Indian J. Pure Appl. Phys. 16, 787 (1978) — P.C. Mishra

p-Benzoquinone-D_4 ($C_6D_4O_2$)

Visible Spectrum — Spectrochim. Acta 20, 1563 (1964) — J.M. Hollas

Emission Spectrum in Vapour — Spectrochim. Acta 21, 1597 (1965) — M.G. Jayswal and R.S. Singh

Absorption Spectrum — Can. J. Chem. 43, 3253 (1965) — J.P. Bryne and I.G. Ross

Absorption Spectrum — J. Mol. Spectrosc. 20, 312 (1966) — I.G. Ross, J.M. Hollas and K.K. Innes

Emission Spectrum in Vapour — Indian J. Pure Appl. Phys. 6, 91 (1968) — R.S. Singh

Calculation — J. Mol. Struct. 14, 75 (1972) — O.F. Bizri and F.F. Prichard

Vibrational Analysis — J. Mol. Spectrosc. 50, 1 (1974) — T.M. Dunn and A.H. Francis

Benzotrichloride ($C_6H_5CCl_3$)

UV Absorption in Vapour — J. Amer. Chem. Soc. 70, 2842 (1948) — W.F. Hanner and F.A. Matsen

Vibrational Analysis — J. Chem. Phys. 42, 35 (1965) — C.V. Stevenson and W.C. Coburn

UV Absorption in Vapour — Appl. Spectrosc. 23, 549 (1969) — K.P.R. Nair, R. Amni Amma and M.P. Srivastava

Benzotrifluoride ($C_6H_5CF_3$)

UV Absorption in Vapour — J. Opt. Soc. Amer. 39 840 (1949) — H. Sponer and D.H. Lowe

Vibrational Analysis — J. Chem. Phys. 27, 740 (1957) — N.A. Narasimham, J.R. Nielsen and R. Theimer

Vibrational Analysis — J. Amer. Chem. Soc.
  81, 1015 (1959) — D.W. Scott, D.R.
  Douslin, J.F. Messely, S.S. Todd, I.A.
  Hossenlopp, T.C. Kinchelve and J.P.
  McCollough
Vibrational Analysis — Indian J. Phys. **36**,
  59 (1962) — K.K. Deb
Fluorescence — J. Chem. Phys. **55**, 5753
  (1971) — D. Gray and D. Phillips
Calculation — J. Mol. Struct. **16**, 365
  (1973) — R.T.C. Brownlee, D.G. Cameron,
  R.D. Topsom, A.R. Katritzky and A.J.
  Sparrow
Vibrational Analysis — Spectrochim. Acta
  33A, 837 (1977) — J.H.S. Green and
  D.J. Harrison

Benzoylbromide (C_6H_5COBr)

IR Spectrum — Spectrochim. Acta **18**, 273
  (1962) — C.N.R. Rao and R. Venkates-
  raghavan
UV Absorption in Vapour — Spectrochim.
  Acta **27A**, 2363 (1971) — M.A. Shashidhar

Benzoychloride (C_6H_5COCl)

Raman — J. Amer. Chem. Soc. **58**, 1953 (1936)—
  D.D. Thompson and J.F. Norris
Raman — Indian J. Phys. **26**, 115 (1952) —
  T.A. Hariharan
IR Spectrum — Spectrochim. Acta **18**, 273
  (1962) — C.N.R. Rao and R. Venkates-
  raghavan
IR Spectrum — J. Chim. Phys. **62**, 3 (1965) —
  D.P. Pierre, V. Lorenzelli and A.
  Alemagna
Vibrational Analysis — Indian J. Phys. **42**,
  610 (1968) — S. Chattopadhyay and J.
  Jha
Vibrational Analysis — Indian J. Phys. **42**,
  243 (1968) — S.C. Sirkar and P.K.
  Bishnui
IR Spectrum — Indian J. Pure Appl. Phys.
  8, 116 (1970) — S.R. Singh, B.B. Lal,
  I.S. Singh and M.P. Srivastava
UV Absorption in Vapour — Indian J. Pure
  Appl. Phys. **8**, 856 (1970) — V.N. Verma,
  K.P.R. Nair and M.P. Srivastava
UV Absorption in Vapour — Spectrochim. Acta
  27A, 2363 (1971) — M.A. Shashidhar
IR Spectrum — Indian J. Phys. **46**, 412
  (1972) — S.C. Bag and G.S. Kastha

Benzoylcyanide (C_6H_5COCN)

UV Absorption in Vapour — Spectrochim.
  Acta **27A**, 2363 (1971) — M.A. Shashidhar
IR Spectrum — Indian J. Pure Appl. Phys.
  9, 138 (1971) — M.A. Shashidhar

Benzoylfluoride (C_6H_5COF)

Vibrational Analysis — Acta Phys. Aust.
  1, 352 (1948) — H. Seewann-Albert and
  L. Kahovec
Vibrational Analysis — Spectrochim. Acta
  33A, 583 (1977) — J.H.S. Green and
  D.J. Harrison

Benzoylhydroxy (C_6H_5COOH)

Vibrational Analysis — Spectrochim. Acta
  33A, 583 (1977) — J.H.S. Green and
  D.J. Harrison

Benzoylmethoxy ($C_6H_5COOCH_3$)

Vibrational Analysis — Spectrochim. Acta
  33A, 583 (1977) — J.H.S. Green and
  D.J. Harrison

Benzoylmethyl ($C_6H_5COCH_3$)

Vibrational Analysis — Spectrochim. Acta
  33A, 583 (1977) — J.H.S. Green and
  D.J. Harrison

Benzylamine ($C_6H_5CH_2NH_2$)

UV Emission Spectrum in Vapour — J. Sci.
  Res. BHU (India) **15(1)**, 97 (1964) —
  N.L. Singh and S.N. Singh
IR Spectrum — Spectrochim. Acta **19**, 243
  (1963) — R. Leysen and J. Van Rysssel-
  berg
Vibrational Analysis — Indian J. Phys.
  41, 759 (1967) — S. Chattopadhyay

N-Benzylamine ($C_6H_5NHCH_2C_6H_5$)

IR Spectrum — Spectrochim. Acta **16**, 117
  (1960) — A.G. Moritz

Benzylbromide ($C_6H_5CH_2Br$)

IR Spectrum — Indian J. Phys. <u>41</u>, 759
(1967) — S. Chattopadhyay
Vibrational Analysis — Spectrochim. Acta
<u>28A</u>, 51 (1972) — L. Verdonek and G.P.
der Kelen

Benzylchloride ($C_6H_5CH_2Cl$)

UV Absorption in Vapour — J. Amer. Chem.
Soc. <u>70</u>, 2842 (1948) — W.F. Hamner and
F.A. Matsen
Raman — Indian J. Phys. <u>25</u>, 131 (1951) —
A.K. Roy
UV Absorption in Different States — Indian
J. Phys. <u>26</u>, 233 (1952) — H.N. Swamy
IR Spectrum — Spectrochim. Acta <u>20</u>, 45
(1964) — J.M. Mannion and T.S. Wang
IR Spectrum — Indian J. Phys. <u>41</u>, 759
(1967) — S. Chattopadhyay
UV Absorption in Vapour — Appl. Spectrosc.
<u>23</u>, 616 (1969) — R. Amni Amma, K.P.R.
Nair and D.K. Rai
Vibrational Analysis — Spectrochim. Acta
<u>28A</u>, 51 (1972) — L. Verdonek and G.P.
Van der Kelen

Benzylcyanide ($C_6H_5CH_2CN$)

IR Spectrum — Indian J. Phys. <u>41</u>, 759
(1967) — S. Chattopadhyay

Benzyliodide ($C_6H_5CH_2I$)

Vibrational Analysis — Spectrochim. Acta
<u>28A</u>, 51 (1972) — L. Verdonek and G.P.
Van der Kelen

Benzylmethyl ($C_6H_5CH_2CH_3$)

Vibrational Analysis — Spectrochim. Acta
<u>28A</u>, 51 (1972) — L. Verdonek and G.P.
Van der Kelen

Biphenyl ($C_6H_5C_6H_5$)

Absorption Spectrum — Proc. Indian Acad.
Soc. <u>3</u>, 148 (1936) — P.K. Seshan
Absorption Spectrum — J. Chem. Phys. <u>4</u>,
760 (1936) — E.P. Carr and H. Stucklen
Absorption Spectrum — Helv. Chim. Acta <u>3</u>,
2092 (1950) — F. Almasy and H. Laemmel

Vibrational Analysis — Spectrochim. Acta
<u>5</u>, 627 (1952) — J.E. Katon and
E.R. Lippincott
Calculation — Opt. Spektrosk. <u>1</u>, 742
(1956) — M.A. Kovner
IR Spectrum— Acta Chem. Scand. <u>11</u>, 640
(1957) — J. Dale
Calculation — Opt. Spektrosk. <u>9</u>, 155
(1960) — G.V. Peregudov
Vibrational Analysis — J. Mol. Spectrosc.
<u>6</u>, 238 (1961) — D. Steele and E.R.
Lippincott
IR Spectrum — J. Mol. Spectrosc. <u>14</u>, 308
(1964) — J.E. Katon and N.T. McDevitt
Vibrational Analysis — Spectrochim. Acta
<u>22</u>, 235 (1966) — S. Scandronic and
F. Geiss
IR Spectrum — Spectrochim. Acta <u>22</u>, 501
(1966) — W.R. McWhinnie and R.C. Poller
IR Spectrum — J. Chim. Phys. <u>64</u>, 765
(1967) — B. Pasquier and J.M. Lebas
Vibrational Analysis — Spectrochim. Acta
<u>24A</u>, 483 (1968) — G. Zerbi and S.
Sandroni
Calculation — Spectrochim. Acta <u>24A</u>, 511
(1968) — G. Zerbi and S. Sandroni
IR Spectrum — Mol. Cryst. Lid. Cryst.
<u>11</u>, 35 (1970) — B. Pasquier
Vibrational Analysis — J. Mol. Struct. <u>11</u>,
105 (1972) — R.M. Barrett and D.
Steele
UV Absorption — J. Chem. Phys. <u>58</u>, 5078
(1973) — R.M. Hochstrasser, R.D.
McAlpine and J.D. Whiteman
Vibrational Analysis — Spectrochim. Acta
<u>31A</u>, 1569 (1975) — A.V. Bree, M.
Edelson and R.A. Kidd

Biphenyl-D_2 ($C_6H_5C_6H_3D_2$)

Vibrational Analysis — Spectrochim. Acta
<u>22</u>, 235 (1966) — S. Sandroni and F.
Geiss
Vibrational Analysis — Spectrochim. Acta
<u>30A</u>, 1731 (1974) — R.M. Barrett and
D. Steele

Biphenyl-D_{10} ($C_6D_5C_6D_5$)

Vibrational Analysis — Opt. Spektrosk. <u>1</u>,
742 (1956) — M.A. Kovner
Vibrational Analysis — Spectrochim. Acta
<u>15</u>, 627 (1959) — J.E. Katon and E.R.
Lippincott
Raman — Opt. Spektrosk. <u>9</u>, 155 (1960) —
G.V. Peregudov

IR Spectrum – J. Mol. Spectrosc. <u>14</u>, 308
 (1964) – J.E. Katon and N.T. McDevitt
Vibrational Analysis – Spectrochim. Acta
 <u>22</u>, 235 (1966) – S. Sandroni and F.
 Geiss
Vibrational Analysis – Spectrochim. Acta
 <u>24A</u>, 483 (1968) – G. Zerbi and S.
 Sandroni
Calculation – Spectrochim. Acta <u>24A</u>, 511
 (1968) – G. Zerbi and S. Sandroni
Vibrational Analysis – J. Chem. Soc.
 Faraday Trans. II <u>69</u>, 1601 (1973) –
 V.J. Eaton and D. Steele
Vibrational Analysis – Spectrochim. Acta
 <u>30A</u>, 1731 (1974) – R.M. Barrett and
 D. Steele
Calculation – Spectrochim. Acta <u>30A</u>, 1731
 (1974) – R.M. Barrett and D. Steele
Vibrational Analysis – Spectrochim. Acta
 <u>31A</u>, 1569 (1975) – A.V. Bree, M. Edel-
 son and R.A. Kidd

o–Bromoaniline $(C_6H_4BrNH_2)$

UV Absorption in Vapour – Indian J. Pure
 Appl. Phys. <u>6</u>, 25 (1968) – G.N.R.
 Tripathi
UV Absorption in Vapour – Indian J. Phys.
 <u>42</u>, 354 (1968) – C.G. Rama Rao
UV Absorption in Vapour – Indian J. Phys.
 <u>43</u>, 557 (1969) – G.N.R. Tripathi

m–Bromoaniline $(C_6H_4BrNH_2)$

Raman – Sber. Akad. Ser Wien. <u>144</u>, 417
 (1935) – K.W.F. Kohlrausch and G.P.
 Ypsilanti
UV Absorption in Vapour – Indian J. Phys.
 <u>42</u>, 354 (1968) – C.G. Rama Rao
UV Absorption in Vapour – Indian J. Pure
 Appl. Phys. <u>6</u>, 25 (1968) – G.N.R.
 Tripathi
UV Absorption in Vapour – Indian J. Phys.
 <u>43</u>, 559 (1969) – G.N.R. Tripathi

p–Bromoaniline $(C_6H_4BrNH_2)$

Raman – Sber Akad. Ser. Wien. <u>144</u>, 417
 (1935) – K.W.F. Kohlrausch and G.P.
 Ypsilanti
UV Absorption in Solution – J. Chem. Soc.
 221 (1949) – J.C. Gage
IR Spectrum – Indian J. Pure Appl. Phys.
 <u>2</u>, 204 (1964) – D. Sharma and L.N.
 Tripathi

UV Absorption in Vapour – Indian J. Pure
 Appl. Phys. <u>2</u>, 204 (1964) – D. Sharma
 and L.N. Tripathi
IR Spectrum – Spectrochim. Acta <u>25A</u>, 1423
 (1969) – L.K. Dyall
UV Absorption in Vapour – Indian J. Pure
 Appl. Phys. <u>8</u>, 730 (1970) – L.N.
 Tripathi
Calculation – J. Mol. Struct. <u>14</u>, 6 (1972)–
 L. Smekankine and J. Etchepare

o–Bromoanisole $(C_6H_4BrOCH_3)$

UV Absorption in Vapour – J. Sci. Ind. Res.
 (India) <u>15B</u>, 261 (1956) – S. Rao and
 V.R.K. Rao
UV Absorption in Different States – Indian
 J. Phys. <u>35</u>, 203 (1961) – T.N. Mishra
 and S.B. Banerjee
Vibrational Analysis – Spectrochim. Acta
 <u>19</u>, 877 (1963) – E.F. Mooney
Vibrational Analysis – Spectrochim. Acta
 <u>25A</u>, 343 (1963) – N.L. Owen and R.E.
 Hester

m–Bromoanisole $(C_6H_4BrOCH_3)$

Vibrational Analysis – Spectrochim. Acta
 <u>19</u>, 877 (1963) – E.F. Mooney
Vibrational Analysis – Spectrochim. Acta
 <u>25A</u>, 343 (1969) – N.L. Owen and R.E.
 Hester
UV Absorption in Vapour – Indian J. Pure
 Appl. Phys. <u>10</u>, 392 (1972) – B.J.
 Ansari and D. Sharma
UV Absorption in Vapour – Indian J. Pure
 Appl. Phys. <u>11</u>, 232 (1973) – D. Marjit
 and S.B. Banerjee

p–Bromoanisole $(C_6H_4BrOCH_3)$

Raman – Monatsh. Chem. <u>72</u>, 244 (1937) –
 O. Paulsen
Raman – Monatsh. Chem. <u>76</u>, 200 (1947) –
 E. Herz, K.W.F. Kohlrausch and R.
 Vogel
UV Absorption in Solution – J. Chem. Soc.
 2310 (1952) – A. Burawoy and J.T.
 Chamberlain
UV Absorption in Vapour – J. Sci. Ind. Res.
 (India) <u>15B</u>, 260 (1956) – V. Suryana-
 rayana and V.R. Rao
Vibrational Analysis – J. Chem. Soc. 2051
 (1959) – A.R. Katritzky and P. Simons
UV Absorption in Solution – Spectrochim.
 Acta <u>17</u>, 545 (1961) – E. Spinner

UV Absorption in Solution — Can. J. Chem.
39, 1131 (1961) — W.F. Forbes
Vibrational Analyis — Spectrochim. Acta
19, 877 (1963) — E.F. Mooney
IR Spectrum — Curr. Sci. 36, 427 (1967) —
V.B. Singh
Vibrational Analysis — Spectrochim. Acta
23A, 1111 (1967) — M. Horak, E.R.
Lippincott and R.K. Khanna
Vibrational Analysis — Spectrochim. Acta
25A, 343 (1969) — N.L. Owen and R.E.
Hester
UV Absorption in Different States — Indian
J. Phys. 47, 205 (1973) — D. Marjit

o-Bromobenzaldehyde (C_6H_4BrCHO)

UV Absorption in Vapour — Indian J. Pure
Appl. Phys. 1, 328 (1963) — R.M.P.
Jaiswal and D. Sharma
UV Absorption in Vapour — Indian J. Pure
Appl. Phys. 2, 232 (1964) — R.M.P.
Jaiswal
Vibrational Analysis — Curr. Sci. 36, 365
(1967) — V.B. Singh and I.S. Singh
IR Spectrum — Indian J. Pure Appl. Phys.
6, 81 (1968) — V.B. Singh and I.S.
Singh
UV Absorption in Vapour — Indian J. Pure
Appl. Phys. 7, 47 (1969) — R.M.P.
Jaiswal
UV Absorption in Vapour — Indian J. Pure
Appl. Phys. 7, 674 (1969) — R.M.P.
Jaiswal
Raman — Indian J. Pure Appl. Phys. 8, 432
(1970) — B. Singh and R.M.P. Jaiswal
Vibrational Analysis — Spectrochim. Acta
32A, 1265 (1976) — J.H.S. Green and
D.J. Harrison

m-Bromobenzaldehyde (C_6H_4BrCHO)

UV Absorption in Vapour — Indian J. Pure
Appl. Phys. 1, 338 (1963) — R.M.P.
Jaiswal and D. Sharma
UV Absorption in Vapour — Indian J. Pure
Appl. Phys. 2, 232 (1964) — R.M.P.
Jaiswal
Vibrational Analysis — Curr. Sci. 36, 365
(1967) — V.B. Singh and I.S. Singh
IR Spectrum — Indian J. Pure Appl. Phys.
6, 81 (1968) — V.B. Singh and I.S.
Singh
UV Absorption in Vapour — Indian J. Pure
Appl. Phys. 7, 47 (1969) — R.M.P.
Jaiswal

Raman — Indian J. Pure Appl. Phys. 8, 432
(1970) — B. Singh and R.M.P. Jaiswal
Vibrational Analysis — Spectrochim. Acta
32A, 1265 (1976) — J.H.S. Green and
D.J. Harrison

p-Bromobenzaldehyde (C_6H_4BrCHO)

Raman — Z. Phys. Chem. B22, 169 (1933) —
G.B. Bonino and R. Manzoni-Ansidei
Raman — Z. Phys. Chem. B38, 119 (1937) —
L. Kahovec and K.W.F. Kohlrausch
UV Absorption in Vapour — Indian J. Pure
Appl. Phys. 1, 338 (1963) — R.M.P.
Jaiswal and D. Sharma
UV Absorption in Vapour — Indian J. Pure
Appl. Phys. 2, 232 (1964) — R.M.P.
Jaiswal
Vibrational Analysis — Curr. Sci. 36, 365
(1967) — V.B. Singh and I.S. Singh
Raman — Spectrochim. Acta 23A, 462 (1967) —
F.B. Brown
IR Spectrum — Indian J. Pure Appl. Phys.
6, 81 (1968) — V.B. Singh and I.S.
Singh
UV Absorption in Vapour — Indian J. Pure
Appl. Phys. 7, 674 (1969) — R.M.P.
Jaiswal
UV Absorption in Vapour — Indian J. Pure
Appl. Phys. 7, 47 (1969) — R.M.P.
Jaiswal
Raman — Indian J. Pure Appl. Phys. 8, 432
(1970) — B. Singh and R.M.P. Jaiswal
Vibrational Analysis — Spectrochim. Acta
32A, 1265 (1976) — J.H.S. Green and
D.J. Harrison

Bromobenzene (C_6H_5Br)

Raman — J. Chem. Phys. 1, 406 (1933) — J.W.
Murray and D.H. Andrews
UV Absorption Spectrum — Proc. Phys. Math.
Soc. Jap. 22, 685 (1940) — K. Asagoe
and Y. Ikemoto
Raman — J. Chem. Phys. 9, 667 (1941) — H.
Sponer and J. Skirby-Smith
UV Absorption in Vapour — Phys. Rev. 59,
924 (1941) — I. Walerstein
Vibrational Analysis — Proc. Indian Acad.
Sci A15, 401 (1942) — C.S. Venkateswa-
ran and N.S. Pandya
UV Absorption in Vapour — Curr. Sci. 18,
437 (1949) — K. Sreeramamurty
Raman — Indian J. Phys. 29, 503 (1955) —
D.C. Biswas
Vibrational Analysis — J. Chem. Soc. 1350
(1956) — D.H. Whiffen

Raman — Indian J. Phys. <u>31</u>, 635 (1957) —
    G.S. Kastha
Vibrational Analysis — J. Amer. Chem. Soc.
    <u>81</u>, 1015 (1959) — D.W. Scott, D.R.
    Douslin, J.F. Messerly, S.S. Todd, I.A.
    Hossenlopp, T.C. Kincheloe and J.P.
    McCollough
UV Absorption in Solution — Can. J. Chem.
    <u>39</u>, 1131 (1961) — W.F. Forbes
UV Absorption in Vapour — J. Sci. Ind. Res.
    (India) <u>21B</u>, 512 (1962) — S. Prakash
    and N.L. Singh
IR Spectrum — Indian J. Phys. <u>38</u>, 610
    (1964) — S.C. Sirkar, D.K. Mukherjee
    and P.K. Bishnui
Vibrational Analysis — Spectrochim. Acta
    <u>21</u>, 1495 (1965) — T.R. Nanney, R.T.
    Bailey and E.R. Lippincott
IR Spectrum — Spectrochim. Acta <u>22</u>, 501
    (1966) — W.R. McWhinnie and R.C. Poller
Vibrational Analysis — Proc. Roy. Soc.
    <u>A298</u>, 51 (1967) — P.R. Griffiths and
    H.W. Thompson
UV Absorption in Vapour — Mol. Phys. <u>19</u>,
    297 (1970) — T. Cvitas
Raman — J. Mol. Struct. <u>5</u>, 477 (1970) —
    N.T. McDevitt and W.G. Fateley
Raman (High Resolution) — J. Mol. Spectrosc.
    <u>39</u>, 73 (1971) — C.T. Meneely, C.Y. She
    and D.F. Edwards
Calculation — J. Mol. Struct. <u>16</u>, 365
    (1973) — R.T.C. Brownlee, D.G. Cameron,
    R.D. Topsom, A.R. Katritzky and A.J.
    Sparrow

Bromobenzene—p—D_1 (C_6H_4BrD)

Vibrational Analysis — Spectrochim. Acta
    <u>33A</u>, 607 (1977) — T. Uno, A. Kuwae
    and K. Machida

Bromobenzene—2,3,5,6—D_4 (C_6HBrD_4)

Vibrational Analysis — Spectrochim. Acta
    <u>33A</u>, 607 (1977) — T. Uno, A. Kuwae and
    K. Machida

Bromobenzene—D_5 (C_6BrD_5)

Vibrational Analysis — Spectrochim. Acta
    <u>22</u>, 737 (1966) — T.R. Nanney, E.R.
    Lippincott and J.C. Hamer
Vibrational Analysis — Spectrochim. Acta
    <u>25A</u>, 507 (1969) — P.N. Gates, K. Radc-
    liffe and D. Steele

UV Absorption in Vapour — Indian J. Pure
    Appl. Phys. <u>10</u>, 635 (1972) — R.S.
    Tripathi and B.R. Pandey

p—Bromobenzenethiol (C_6H_4BrSH)

Vibrational Analysis — Spectrochim. Acta
    <u>12</u>, 305 (1958) — C. Garrigau—Lagrange,
    J.M. Lebas and M.L. Josien
Vibrational Analysis — Spectrochim. Acta
    <u>26A</u>, 1515 (1970) — J.H.S. Green, D.J.
    Harrison, W. Kynaston and D.W. Scott

o—Bromobenzoic acid ($C_6H_4CO_2HBr$)

Vibrational Analysis — Spectrochim. Acta
    <u>33A</u>, 575 (1977) — J.H.S. Green

m—Bromobenzoic acid ($C_6H_4CO_2HBr$)

Vibrational Analysis — Spectrochim. Acta
    <u>33A</u>, 575 (1977) — J.H.S. Green

p—Bromobenzoic acid ($C_6H_4CO_2HBr$)

Vibrational Analysis — Spectrochim. Acta
    <u>33A</u>, 575 (1977) — J.H.S. Green

o—Bromobenzonitrile (C_6H_4BrCN)

UV Absorption in Vapour — Indian J. Pure
    Appl. Phys. <u>14</u>, 587 (1976) — S.M.
    Pandey and S.J. Singh
Vibrational Analysis — Spectrochim. Acta
    <u>32A</u>, 1279 (1976) — J.H.S. Green and
    D.J. Harrison
Vibrational Analysis — Spectrochim. Acta
    <u>33A</u>, 837 (1977) — J.H.S. Green and
    D.J. Harrison

m—Bromobenzonitrile (C_6H_4BrCN)

Vibrational Analysis — Spectrochim. Acta
    <u>32A</u>, 1279 (1976) — J.H.S. Green and
    D.J. Harrison
Vibrational Analysis — Spectrochim. Acta
    <u>33A</u>, 837 (1977) — J.H.S. Green and
    D.J. Harrison

p-Bromobenzonitrile (C_6H_4BrCN)

Raman — Sber. Akad. Wiss. Wien, 144, 285, 417 (1935) — K.W.F. Kohlrausch and G.P. Ypsilanti

Raman — Monatsh. Chem. 66, 385 (1935) — K.W.F. Kohlrausch and G.P. Ypsilanti

Vibrational Analysis — Z. Elektrochem. 64, 1228 (1960) — S. Weckherlin and W. Luttke

Vibrational Analysis — Spectrochim. Acta 21, 45 (1965) — H.W. Wilson and J.E. Bloor

UV Absorption in Vapour — Indian J. Pure Appl. Phys. 4, 169 (1966) — S.M. Pandey and B.R. Pandey

IR Spectrum — Indian J. Pure Appl. Phys. 4, 169 (1966) — S.M. Pandey and B.R. Pandey

IR Spectrum — Indian J. Phys. 42, 33 (1968) — B.R. Pandey

UV Absorption in Vapour — Indian J. Phys. 42, 33 (1968) — B.R. Pandey

Vibrational Analysis — Spectrochim. Acta 32A, 1279 (1976) — J.H.S. Green and D.J. Harrison

Vibrational Analysis — Spectrochim. Acta 33A, 837 (1977) — J.H.S. Green and D.J. Harrison

p-Bromobenzylbromide ($C_6H_4BrCH_2Br$)

Vibrational Analysis — Spectrochim. Acta 28A, 55 (1972) — L. Verdonck and G.P. Van der Kelen

o-Bromochlorobenzene (C_6H_4BrCl)

Raman — Monatsh. Chem. 65, 199 (1935) — K.W.F. Kohlrausch and A. Pongratz

UV Absorption in Solution — Z. Phys. Chem. 33, 311 (1936) — H. Conrad-Billroth and G. Forster

UV Absorption in Vapour — Indian J. Phys. 31, 387 (1957) — S.L.N.G. Krishnamachari

Vibrational Analysis — Curr. Sci. 26, 144 (1957) — S.L.N.G. Krishnamachari

UV Absorption in Different States — Indian J. Pure Appl. Phys. 7, 62 (1969) — K.G. Srinivasacharya

Vibrational Analysis — Spectrochim. Acta 26A, 1913 (1970) — J.H.S. Green

m-Bromochlorobenzene (C_6H_4BrCl)

UV Absorption in Solution — Z. Phys. Chem. 33, 311 (1936) — H. Conrad-Billroth and G. Forster

Vibrational Analysis — Curr. Sci. 26, 144 (1957) — S.L.N.G. Krishnamachari

UV Absorption in Vapour — Indian J. Phys. 31, 387 (1957) — S.L.N.G. Krishnamachari

Vibrational Analysis — Opt. Spektrosk. 5, 134 (1958) — M.A. Kovner and G.V. Peregudov

UV Absorption in Solid — Indian J. Pure Appl. Phys. 2, 202 (1964) — K.G. Srinivasacharya and C. Santhama

UV Absorption in Different States — Indian J. Pure Appl. Phys. 7, 62 (1969) — K.G. Srinivasacharya

Vibrational Analysis — Spectrochim. Acta 26A, 1523 (1970) — J.H.S. Green

p-Bromochlorobenzene (C_6H_4BrCl)

UV Absorption in Solution — Z. Phys. Chem. 33, 311 (1936) — H. Conrad-Billroth and G. Forster

UV Absorption in Vapour — Bull. Sec. Sci. Akad. Rou. 23, 34 (1940) — G.A. Dima and H. Tintea

UV Absorption in Vapour — Indian J. Phys. 31, 387 (1957) — S.L.N.G. Krishnamachari

Vibrational Analysis — Spectrochim. Acta 12, 57 (1958) — A. Stojilkovic and D.H. Whiffen

Vibrational Analysis — Spectrochim. Acta 12, 305 (1958) — C. Garrigou-Lagrange, J.M. Lebas and M.L. Josien

Vibrational Analysis — Spectrochim. Acta 13, 225 (1959) — J.M. Lebas, C. Garrigou Lagrange and M.L. Josien

UV Absorption in Solid — Indian J. Pure Appl. Phys. 2, 202 (1964) — K.G. Srinivasacharya and C. Santhama

Vibrational Analysis — Proc. Roy. Soc. A298, 51 (1967) — P.R. Griffiths and H.W. Thompson

Vibrational Analysis — J. Chim. Phys. 64, 1450 (1967) — P. Pommez, M. Lafaix, P. Delorme and V. Lorenzelli

UV Absorption in Different States — Indian J. Pure Appl. Phys. 7, 62 (1969) — K.G. Srinivasacharya

Vibrational Analysis — Spectrochim. Acta 26A, 1503 (1970) — J.H.S. Green

Vibrational Analysis (Crystal) — Indian J. Phys. 51B, 50 (1977) — Shyampati, I.S. Singh and J. Shamir

2-Bromo-4-Chlorophenol ($C_6H_3BrClOH$)

UV Absorption in Vapour – Indian J. Pure
Appl. Phys. 12, 296 (1974) – P.K.
Mallick and S.B. Banerjee
Vibrational Analysis – Indian J. Pure Appl.
Phys. 12, 296 (1974) – P.K. Mallick
and S.B. Banerjee

4-Bromo-2-Chlorophenol ($C_6H_3BrClOH$)

UV Absorption in Vapour – Indian J. Pure
Appl. Phys. 12, 296 (1974) – P.K.
Mallick and S.B. Banerjee
Vibrational Analysis – Indian J. Pure Appl.
Phys. 12, 296 (1974) – P.K. Mallick
and S.B. Banerjee

3-Bromo-4-Chlorotoluene ($C_6H_3BrClCH_3$)

Vibrational Analysis – Bull. Soc. Chim.
France 13, 540 (1946) – R. Pajeau

5-Bromo-2-Chlorotoluene ($C_6H_3BrClCH_3$)

Vibrational Analysis – Bull. Soc. Chim.
France 13, 540 (1946) – R. Pajeau

1-Bromo-2,3-Dimethylbenzene ($C_6H_3Br(CH_3)_2$)

IR Spectrum – Indian J. Phys. 51A, 88
(1977) – R.N. Singh and K.N. Upadhya
UV Absorption in Vapour – Indian J. Pure
Appl. Phys. 16, 721 (1978) – P.V.
Shanbhag, M.A. Shashidhar and K.S. Rao
Vibrational Analysis – Spectrochim. Acta
34A, 39 (1978) – R.N. Singh, S.C. Prasad
and R.K. Prasad
UV Absorption in Vapour – Indian J. Phys.
52B, 225 (1978) – P.V. Shanbhag, M.A.
Shashidhar and K.S. Rao

1-Bromo-2,4-Dimethylbenzene ($C_6H_3Br(CH_3)_2$)

IR Spectrum – Indian J. Phys. 51A, 88
(1977) – R.N. Singh and K.N. Upadhya
Vibrational Analysis – Indian J. Pure Appl.
Phys. 16, 64 (1978) – R.N. Singh, S.C.
Prasad and R.K. Prasad
UV Absorption in Vapour – Indian J. Pure
Appl. Phys. 16, 721 (1978) – P.V.
Shanbhag, M.A. Shashidhar and K.S. Rao

UV Absorption in Vapour – Indian J. Phys.
52B, 225 (1978) – P.V. Shanbhag, M.A.
Shashidhar and K.S. Rao
Vibrational Analysis – Spectrochim. Acta
34A, 39 (1978) – R.N. Singh, S.C.
Prasad and R.K. Prasad

1-Bromo-2,6-Dimethylbenzene ($C_6H_3Br(CH_3)_2$)

IR Spectrum – Indian J. Phys. 51A, 88
(1977) – R.N. Singh and K.N. Upadhya
Vibrational Analysis – Indian J. Pure Appl.
Phys. 15, 264 (1977) – R.N. Singh
and S.C. Prasad
UV Absorption in Vapour – Indian J. Pure
Appl. Phys. 16, 721 (1978) – P.V.
Shanbhag, M.A. Shashidhar and K.S. Rao
UV Absorption in Vapour – Indian J. Phys.
52B, 225 (1978) – P.V. Shanbhag, M.A.
Shashidhar and K.S. Rao
Vibrational Analysis – Spectrochim. Acta
34A, 39 (1978) – R.N. Singh, S.C.
Prasad and R.K. Prasad

1-Bromo-3,5-Dimethylbenzene ($C_6H_3Br(CH_3)_2$)

Raman – Monatsh. Chem. 76, 215 (1946) –
K.W.F. Kohlrausch
Vibrational Analysis – Spectrochim. Acta
27A, 793 (1971) – J.H.S. Green, D.J.
Harrison and W. Kynaston

1-Bromo-2,4-Dinitrobenzene ($C_6H_3Br(NO_2)_2$)

Vibrational Analysis – Spectrochim. Acta
20, 1021 (1964) – E.F. Mooney

2-Bromoethylbenzene ($C_6H_4BrC_2H_5$)

UV Absorption in Vapour – Indian J. Pure
Appl. Phys. 8, 182 (1970) – K.G.
Srinivasacharya

Bromoethynylbenzene (C_6H_5CCBr)

Vibrational Analysis – Spectrochim. Acta
26A, 919 (1970) – R.D. McLachlan

28

o-Bromofluorobenzene (C_6H_4BrF)

UV Absorption in Solution — Z. Phys. Chem.
33, 311 (1936) — H. Conrad-Billroth
and G. Forster

UV Absorption in Vapour — Bull. Sec. Sci.
Akad. Rou. 23, 34 (1940) — G.A. Dima
and H. Tintea

Raman — Curr. Sci. 25, 185 (1956) — S.L.N.
G. Krishnamachari

IR Spectrum — Curr. Sci. 25, 355 (1956) —
S.L.N.G. Krishanamachari

UV Absorption in Vapour — Indian J. Phys.
30, 487 (1956) — S.L.N.G. Krishnamachari

Vibrational Analysis — Curr. Sci. 26, 144
(1957) — S.L.N.G. Krishnamachari

Vibrational Analysis — Spectrochim. Acta
19, 683 (1963) — G. Varsanyi, S. Holly
and T. Farago

Vibrational Analysis — Spectrochim. Acta
19, 669 (1963) — G. Varsanyi, S. Holly
and T. Farago

Calculation — Curr. Sci. 35, 331 (1966) —
N.A. Narasimhan and C.V.S.R. Rao

Normal Coordinate Analysis — J. Mol.
Spectrosc. 28, 44 (1968) — N.A. Narasimhan and C.V.S.R. Rao

Calculation — J. Mol. Spectrosc. 30, 192
(1969) — N.A. Narasimhan and C.V.S.R.
Rao

Vibrational Analysis — Spectrochim. Acta
26A, 1913 (1970) — J.H.S. Green

o-Bromofluorobenzene-D_4 (C_6D_4BrF)

Vibrational Analysis — J. Mol. Spectrosc.
48, 185 (1973) — N.D. Patel, V.B.
Kartha and N.A. Narasimhan

m-Bromofluorobenzene (C_6H_4BrF)

UV Absorption in Solution — Z. Phys. Chem.
33, 311 (1936) — H. Conrad-Billroth
and G. Forster

UV Absorption in Vapour — Bull. Sec. Sci.
Akad. Rou. 23, 34 (1940) — G.A. Dima
and H. Tintea

Raman — Monatsh. Chem. 76, 249 (1947) —
K.W.F. Kohlrausch

UV Absorption in Vapour — Indian J. Phys.
30, 487 (1956) — S.L.N.G. Krishnamachari

Raman — Curr. Sci. 25, 260 (1956) — S.L.N.
G. Krishnamachari

Vibrational Analysis — Curr. Sci. 26, 144
(1957) — S.L.N.G. Krishnamachari

Vibrational Analysis — Spectrochim. Acta
19, 675 (1963) — G. Varsanyi, S. Holly
and T. Farago

Vibrational Analysis — Spectrochim. Acta
19, 669 (1963) — G. Varsanyi, S. Holly
and T. Farago

Calculation — Curr. Sci. 35, 331 (1966) —
N.A. Narasimhan and C.V.S.R. Rao

Normal Coordinate Analysis — J. Mol.
Spectrosc. 28, 44 (1968) — N.A. Narasimhan and C.V.S.R. Rao

Calculation — J. Mol. Spectrosc. 30, 192
(1969) — N.A. Narasimhan and C.V.S.R.
Rao

Vibrational Analysis — Spectrochim. Acta
26A, 1523 (1970) — J.H.S. Green

m-Bromofluorobenzene-D_4 (C_6D_4BrF)

Vibrational Analysis — J. Mol. Spectrosc.
48, 185 (1973) — N.D. Patel, V.B.
Kartha and N.A. Narasimhan

p-Bromofluorobenzene (C_6H_4BrF)

UV Absorption in Solution — Z. Phys. Chem.
33, 311 (1936) — H. Conrad-Billroth
and G. Forster

UV Absorption in Vapour — Bull. Sec. Sci.
Akad. Rou. 23, 34 (1940) — G.A. Dima
and H. Tintea

UV Absorption in Vapour — Indian J. Phys.
30, 487 (1956) — S.L.N.G. Krishnamachari

Vibrational Analysis — J. Chem. Phys. 24,
420 (1956) — N.A. Narasimhan, M.Z. El
Sabban and J.R. Nielsen

Vibrational Analysis — J. Chem. Phys. 24,
433 (1956) — N.A. Narasimhan and J.R.
Nielsen

Vibrational Analysis — Spectrochim. Acta
12, 305 (1958) — C. Garrigou-Lagrange,
J.M. Lebas and M.L. Josien

Vibrational Analysis — Spectrochim. Acta
12, 57 (1958) — A. Stojilkovic and
D.H. Whiffen

Vibrational Analysis — Spectrochim. Acta
19, 669 (1963) — G. Varsanyi, S. Holly
and T. Farago

Calculation — Curr. Sci. 35, 331 (1966) —
N.A. Narasimhan and C.V.S.R. Rao

Normal Coordinate Analysis — J. Mol.
Spectrosc. 23, 44 (1967) — N.A. Narasimhan and C.V.S.R. Rao

Vibrational Analysis — Proc. Roy. Soc.
A298, 51 (1967) — P.R. Griffiths and
H.W. Thompson

Vibrational Analysis — J. Chim. Phys. 64, 1450 (1967) — P. Pommez, M. Lafaix, P. Delorme and V. Lorenzelli
Normal Coordinate Analysis — J. Mol. Spectrosc. 30, 192 (1969) — N.A. Narasimhan and C.V.S.R. Rao
Vibrational Analysis — Spectrochim. Acta 26A, 1503 (1970) — J.H.S. Green

p-Bromofluorobenzene-D_4 (C_6D_4BrF)

Vibrational Analysis — J. Mol. Spectrosc. 48, 185 (1973) — N.D. Patel, V.B. Kartha and N.A. Narasimhan

o-Bromoiodobenzene (C_6H_4BrI)

Raman — J. Chim. Phys. 62, 347 (1965) — M. Brigodiot and J.M. Lebas
Vibrational Analysis — Spectrochim. Acta 26A, 1913 (1970) — J.H.S. Green

m-Bromoiodobenzene (C_6H_4BrI)

Vibrational Analysis — Spectrochim. Acta 26A, 1523 (1970) — J.H.S. Green

p-Bromoiodobenzene (C_6H_4BrI)

Vibrational Analysis — Spectrochim. Acta 12, 305 (1958) — C. Garrigou-Lagrange, J.M. Lebas and M.L. Josien
Vibrational Analysis — Spectrochim. Acta 12, 57 (1958) — A. Stojilkovic and D.H. Whiffen
Vibrational Analysis — Proc. Roy. Soc. A298, 51 (1967) — P.R. Griffiths and H.W. Thompson
Vibrational Analysis — J. Chim. Phys. 64, 1450 (1967) — P. Pommez, M. Lafaix, P. Delorme and V. Lorenzelli
UV Absorption in Different States — Indian J. Pure Appl. Phys. 7, 62 (1969) — K.G. Srinivasacharya
Vibrational Analysis — Spectrochim. Acta 26A, 1503 (1970) — A. Stojilkovic and D.H. Whiffen

3-Bromo-4-Iodotoluene ($C_6H_3BrICH_3$)

Vibrational Analysis — Bull. Soc. Chim. France 13, 540 (1946) — R. Pajeau

1-Bromonaphthalene ($C_{10}H_7Br$)

Raman — Z. Elektrochem. 52, 210 (1948) — V.H. Luther
Phosphorescence — J. Chem. Soc. 3160 (1954) — J. Ferguson, T. Iredale and J.A. Taylor
UV Absorption in Solution — J. Chem. Soc. 304 (1954) — J. Ferguson
UV Absorption in Different States — Indian J. Phys. 28, 121 (1954) — A.R. Deb
IR Spectrum — J. Chem. Soc. 3645 (1954) — J. Ferguson and R.L. Werner
Vibrational Analysis — Indian J. Pure Appl. Phys. 10, 885 (1972) — O.P. Sharma and R.D. Singh
UV Absorption in Vapour — Indian J. Pure Appl. Phys. 14, 583 (1976) — R.D. Singh and S.N. Singh
Vibrational Analysis — Spectrochim. Acta 34A, 985 (1978) — S.N. Singh, H.S. Bhatti and R.D. Singh

2-Bromonaphthalene ($C_{10}H_7Br$)

Raman — Z. Elektrochem. 52, 210 (1948) — V.H. Luther
UV Absorption in Solution — J. Chem. Soc. 304 (1954) — J. Ferguson
Phosphorescence — J. Chem. Soc. 3160 (1954) — J. Ferguson, T. Iredale and J.A. Taylor
IR Spectrum — J. Chem. Soc. 3645 (1954) — J. Ferguson and R.L. Werner
Vibrational Analysis — Indian J. Pure Appl. Phys. 11, 618 (1973) — O.P. Sharma and R.D. Singh

2-Bromo-4-Nitroaniline ($C_6H_3BrNO_2NH_2$)

IR Spectrum — Aust. J. Chem. 23, 947 (1970) — L.K. Dyall

o-Bromonitrobenzene ($C_6H_4BrNO_2$)

Raman — Z. Phys. Chem. 52B, 315 (1942) — H. Wittek
Vibrational Analysis — Spectrochim. Acta 20, 1021 (1964) — E.F. Mooney
Vibrational Analysis — J. Chim. Phys. 62, 347 (1965) — M. Brigodiot and J.M. Lebas
Vibrational Analysis — Spectrochim. Acta 26A, 1925 (1970) — J.H.S. Green and D.J. Harrison

m-Bromonitrobenzene ($C_6H_4BrNO_2$)

Raman — Z. Phys. Chem. 52B, 315 (1942) —
    H. Wittek
Vibrational Analysis — Spectrochim. Acta
    20, 1021 (1964) — E.F. Mooney
Vibrational Analysis — J. Chim. Phys. 63,
    552 (1966) — C. Garrigou-Lagrange,
    M. Chehata and J. Lascombe
Vibrational Analysis — Spectrochim. Acta
    26A, 1925 (1970) — J.H.S. Green and
    D.J. Harrison

p-Bromonitrobenzene ($C_6H_4BrNO_2$)

Raman — Z. Phys. Chem. 52B, 315 (1942) —
    H. Wittek
IR Spectrum — J. Amer. Chem. Soc. 78, 4225
    (1956) — R.D. Kross and V.A. Fassel
Vibrational Analysis — Spectrochim. Acta
    20, 1021 (1964) — E.F. Mooney
Vibrational Analysis — Proc. Roy. Soc.
    A298, 51 (1967) — P.R. Griffiths and
    H.W. Thompson
Vibrational Analysis — J. Chim. Phys. 64,
    1450 (1967) — P. Pommez, M. Lafaix,
    P. Delorme and V. Lorenzelli
Vibrational Analysis — Spectrochim. Acta
    26A, 1503 (1970) — J.H.S. Green
Vibrational Analysis — Spectrochim. Acta
    26A, 1925 (1970) — J.H.S. Green and
    D.J. Harrison
Calculation — J. Mol. Struct. 14, 61
    (1972) — L.Smetankine and J. Etchepare

2-Bromo-4-Nitrotoluene($C_6H_3BrNO_2CH_3$)

IR Spectrum — Indian J. Pure Appl. Phys.
    16, 37 (1978) — M. Rangacharyulu and
    D. Premaswarup

2-Bromo-5-Nitrotoluene($C_6H_3BrNO_2CH_3$)

IR Spectrum — Indian J. Pure Appl. Phys.
    16, 37 (1978) — M. Rangacharyulu and
    D. Premaswarup

Bromopentafluorobenzene (C_6BrF_5)

Vibrational Analysis — Spectrochim. Acta
    19, 1955 (1963) — D.A. Long and D.
    Steele
Vibrational Analysis — Spectrochim. Acta
    22, 695 (1966) — I.J. Hyams, E.R.
    Lippincott and R.T. Bailey

Vibrational Analysis — Spectrochim. Acta
    31A, 1839 (1975) — S.G. Frankiss and
    D.J. Harrison

β-Bromophenetole ($C_6H_5OC_2H_4Br$)

UV Absorption in Vapour — Indian J. Pure
    Appl. Phys. 6, 617 (1968) — Y.P.
    Srivastava, D. Sharma and N.K. Sanyal
IR Spectrum — Indian J. Pure Appl. Phys.
    6, 617 (1968) — Y.P. Srivastava, D.
    Sharma and N.K. Sanyal

o-Bromophenetole ($C_6H_4BrOC_2H_5$)

IR Spectrum — J. Amer. Chem. Soc. 58, 94
    (1936) — L. Pauling
Vibrational Analysis — Spectrochim. Acta
    19, 877 (1963) — E.F. Mooney

m-Bromophenetole ($C_6H_4BrOC_2H_5$)

Vibrational Analysis — Spectrochim. Acta
    19, 877 (1963) — E. F. Mooney

p-Bromophenetole ($C_6H_4BrOC_2H_5$)

Vibrational Analysis — Spectrochim. Acta
    19, 877 (1963) — E.F. Mooney

o-Bromophenol (C_6H_4BrOH)

IR Spectrum — J. Phys. Radium 9, 13
    (1938) — J. Lecomte
IR Spectrum — Trans. Faraday Soc. 36, 333
    (1940) — M.M. Davies
UV Absorption in Solution — Indian J.
    Phys. 30, 353 (1956) — S.B. Banerjee
IR Spectrum — Indian J. Phys. 35, 151
    (1961) — S.B. Banerjee and A.K.
    Chakravarty
IR Spectrum — Spectrochim. Acta 18, 1593
    (1962) — I. Brown, G. Eglinton and
    M. Martin-Smith
IR Spectrum — Spectrochim. Acta 19, 463
    (1963) — I. Brown, G. Eglinton and
    M. Martin-Smith
Vibrational Analysis — Indian J. Pure
    Appl. Phys. 4, 214 (1966) — K. Chandra
UV Absorption in Vapour — Indian J. Pure
    Appl. Phys. 4, 213 (1966) — K. Chandra
Vibrational Analysis — Spectrochim. Acta
    27A, 2199 (1971) — J.H.S. Green, D.J.
    Harrison and W. Kynaston

m-Bromophenol (C_6H_4BrOH)

UV Absorption in Vapour — Indian J. Pure
   Appl. Phys. <u>4</u>, 213 (1966) — K. Chandra
Vibrational Analysis — Indian J. Pure Appl.
   Phys. <u>4</u>, 214 (1966) — K. Chandra
Vibrational Analysis — Spectrochim. Acta
   <u>27A</u>, 2199 (1971) — J.H.S. Green, D.J.
   Harrison and W. Kynaston

p-Bromophenol (C_6H_4BrOH)

Raman — Monatsh. Chem. <u>66</u>, 285 (1935) —
   K.W.F. Kohlrausch and G.P. Ypsilanti
Raman — Sber. Akad. Wiss Wien <u>144</u>, 417
   (1935) — K.W.F. Kohlrausch and G.P.
   Ypsilanti
IR Spectrum — J. Phys. Radium <u>9</u>, 13
   (1938) — J. Lecomte
IR Spectrum — C.R. Akad. Sci. (Paris) <u>212</u>,
   1138 (1941) — M.M. Parodi and M.A.
   Cotton
Vibrational Analysis — Spectrochim. Acta
   <u>13</u>, 225 (1959) — J.M. Lebas, C.
   Garrigou-Lagrange and M.L. Josien
Vibrational Analysis — Appl. Spectrosc.
   <u>16</u>, 32 (1962) — R.J. Jakobsen and E.J.
   Brewer
UV Absorption in Vapour — Indian J. Pure
   Appl. Phys. <u>2</u>, 134 (1964) — K. Chandra
UV Absorption in Vapour — Indian J. Phys.
   <u>39</u>, 464 (1965) — K. Chandra
Vibrational Analysis — Indian J. Pure Appl.
   Phys. <u>4</u>, 214 (1966) — K. Chandra
Vibrational Analysis — Spectrochim. Acta
   <u>27A</u>, 2199 (1971) — J.H.S. Green, D.J.
   Harrison and W. Kynaston

o-Bromopyridine (C_5H_4NBr)

UV Absorption in Solution — J. Chem. Phys.
   <u>22</u>, 1077 (1954) — H.P. Stephenson
UV Absorption in Different States — Indian
   J. Phys. <u>34</u>, 381 (1960) — T.N. Misra
Vibrational Analysis — Spectrochim. Acta
   <u>19</u>, 549 (1963) — J.H.S. Green, W.
   Kynaston and H.M. Paisley

m-Bromopyridine (C_5H_4NBr)

UV Absorption in Solution — J. Chem. Phys.
   <u>22</u>, 1077 (1954) — H.P. Stephenson
UV Absorption in Different States — Indian
   J. Phys. <u>34</u>, 381 (1960) — T.N. Misra
Vibrational Analysis — Spectrochim. Acta
   <u>19</u>, 549 (1963) — J.H.S. Green, W.
   Kynaston and H.M. Paisley

p-Bromopyridine (C_5H_4NBr)

Vibrational Analysis — Spectrochim. Acta
   <u>19</u>, 549 (1963) — J.H.S. Green, W.
   Kynaston and H.M. Paisley

o-Bromopyrimidine ($C_4H_3N_2Br$)

Vibrational Analysis — Spectrochim. Acta
   <u>33A</u>, 189 (1977) — E. Allenstein, P.
   Kiemle, J. Weidlein and W. Podszun

m-Bromoquinoline (C_9H_6NBr)

UV Absorption in Vapour — Indian J. Pure
   Appl. Phys. <u>11</u>, 371 (1973) — M.A.
   Shashidhar

β-Bromostyrene ($C_6H_5CHCHBr$)

IR Spectrum — Curr. Sci. <u>37</u>, 1525 (1968) —
   K. Singh and V.B. Singh
UV Absorption in Vapour — Indian J. Phys.
   <u>42</u>, 668 (1968) — K. Singh and V.B.
   Singh

o-Bromostyrene ($C_6H_4BrCHCH_2$)

Vibrational Analysis — Appl. Spectrosc.
   <u>22</u>, 650 (1968) — W.G. Fateley, G.L.
   Garlson and F.E. Dickson
IR Spectrum — Indian J. Pure Appl. Phys.
   <u>6</u>, 89 (1968) — B.J. Ansari
UV Absorption in Vapour — Indian J. Pure
   Appl. Phys. <u>6</u>, 49 (1968) — B.J. Ansari
Raman — Indian J. Pure Appl. Phys. <u>9</u>, 64
   (1971) — B. Singh and R.M.P. Jaiswal

m-Bromostyrene ($C_6H_4BrCHCH_2$)

Vibrational Analysis — Appl. Spectrosc. <u>22</u>,
   650 (1968) — W.G. Fateley, G.L. Garlson
   and F.E. Dickson
IR Spectrum — Indian J. Pure Appl. Phys.
   <u>6</u>, 89 (1968) — B.J. Ansari
UV Absorption in Vapour — Indian J. Pure
   Appl. Phys. <u>6</u>, 49 (1968) — B.J. Ansari
Raman — Indian J. Pure Appl. Phys. <u>9</u>, 64
   (1971) — B. Singh and R.M.P. Jaiswal

p-Bromostyrene ($C_6H_4BrCHCH_2$)

IR Spectrum — Indian J. Pure Appl. Phys.
   <u>6</u>, 89 (1968) — B.J. Ansari

UV Absorption in Vapour — Indian J. Pure
    Appl. Phys. 6, 49 (1968) — B.J. Ansari
Raman — Indian J. Pure Appl. Phys. 9, 64
    (1971) — B. Singh and R.M.P. Jaiswal

p-Bromothiophenol (C_6H_4BrSH)

Vibrational Analysis — Spectrochim. Acta
    12, 305 (1958) — C. Garrigou-Lagrange,
    J.M. Lebas and M.L. Josien
Vibrational Analysis — Spectrochim. Acta
    13, 225 (1959) — J.M. Lebas, C. Garri-
    gou-Lagrange and M.L. Josien
Vibrational Analysis — Spectrochim. Acta
    26A, 1925 (1970) — J.H.S. Green and
    D.J. Harrison

o-Bromotoluene ($C_6H_4BrCH_3$)

Raman — Monatsh. Chem. 72, 244 (1939) — E.
    Herz
Raman — Indian J. Phys. 28, 423 (1954) —
    D.C. Biswas
Fluorescence — Indian J. Phys. 28, 423
    (1954) — D.C. Biswas
Fluorescence — Indian J. Phys. 30, 407
    (1956) — D.C. Biswas
UV Absorption in Vapour — Indian J. Phys.
    30, 590 (1956) — S.B. Roy
Vibrational Analysis — J. Chem. Soc. 3670
    (1959) — A.R. Katritzky and P. Simons
IR Spectrum — J. Chem. Soc. 2421 (1960) —
    A.R. Katritzky and J.M. Lagowski
UV Absorption in Liquid — Indian J. Phys.
    34, 331 (1960) — J.K. Roy
UV Absorption in Vapour — Indian J. Phys.
    34, 581 (1960) — S.C. Sirkar and J.K.
    Roy
UV Absorption in Solution — Indian J. Phys.
    36, 156 (1962) — J.K. Roy
Vibrational Analysis — Spectrochim. Acta
    20, 1343 (1964) — E.F. Mooney
IR Spectrum — Spectrochim. Acta 22, 1501
    (1966) — G. Joshi and N.L. Singh
Vibrational Analysis — Spectrochim. Acta
    26A, 1913 (1970) — J.H.S. Green

m-Bromotoluene ($C_6H_4BrCH_3$)

Fluorescence — Indian J. Phys. 29, 257
    (1955) — D.C. Biswas
Raman — Indian J. Phys. 29, 257 (1955) —
    D.C. Biswas
UV Absorption in Different States — Indian
    J. Phys. 30, 590 (1956) — S.B. Roy

Fluorescence — Indian J. Phys. 30, 565
    (1956) — D.C. Biswas
UV Absorption in Different States — Indian
    J. Phys. 31, 99 (1957) — S.K. Sen
Vibrational Analysis — J. Chem. Soc. 2058
    (1959) — A.R. Katritzky and P. Simons
UV Absorption in Vapour — Indian J. Phys.
    35, 143 (1961) — J.K. Roy
Vibrational Analysis — Spectrochim. Acta
    20, 1343 (1964) — E.F. Mooney
Vibrational Analysis — Spectrochim. Acta
    26A, 1523 (1970) — J.H.S. Green
Vibrational Analysis — Spectrochim. Acta
    29A, 813 (1973) — L. Verdonck, G.P.
    Van der Kelen and Z. Eeckhaut

p-Bromotoluene ($C_6H_4BrCH_3$)

Raman — Monatsh. Chem. 72, 244 (1939) —
    O. Paulsen
Raman — Indian J. Phys. 28, 423 (1954) —
    D.C. Biswas
Fluorescence — Indian J. Phys. 28, 423
    (1954) — D.C. Biswas
UV Absorption in Different States — Indian
    J. Phys. 30, 590 (1956) — S.B. Roy
Fluorescence — Indian J. Phys. 30, 407
    (1956) — D.C. Biswas
Vibrational Analysis — Spectrochim. Acta
    12, 305 (1958) — C. Garrigou-Lagrange,
    J.M. Lebas and M.L. Josien
Vibrational Analysis — J. Chem. Soc. 2051
    (1959) — A.R. Katritzky and P. Simons
UV Absorption in Vapour — Indian J. Phys.
    35, 143 (1961) — J.K. Roy
UV Absorption in Solid — Indian J. Phys.
    37, 252 (1963) — R.R. Paul and S.C.
    Sirkar
Vibrational Analysis — Spectrochim. Acta
    20, 1343 (1964) — E.F. Mooney
UV Absorption in Vapour — Indian J. Pure
    Appl. Phys. 4, 83 (1966) — G. Joshi
Vibrational Analysis — Spectrochim. Acta
    26A, 1503 (1970) — J.H.S. Green

Bromoxylenes ($C_6H_3Br(CH_3)_2$)

See   Bromodimethylbenzenes

Chloranil ($C_6Cl_4O_2$)

UV Absorption in Solution — J. Chem. Soc.
    44 &490 (1945) — E.A. Braude
UV Absorption in Vapour — Indian J. Pure
    Appl. Phys. 5, 284 (1967) — S.C.
    Srivastava and R.S. Singh

UV Absorption in Liquid — J. Mol. Spectrosc. $\underline{23}$, 191 (1967) — P.E. Stevenson

o-Chloroaniline ($C_6H_4NH_2Cl$)

Raman — Monatsh. Chem. $\underline{74}$, 1 (1943) — E. Herz

Vibrational Analysis — J. Chem. Soc. 3670 (1959) — A.R. Katritzky and R.A. Jones

UV Absorption in Vapour — Indian J. Phys. $\underline{31}$, 577 (1957) — P.B.V. Harnath and K. Sreeramamurty

IR Spectrum — Indian J. Phys. $\underline{37}$, 275 (1963) — K.C. Medhi and G.S. Kastha

Vibrational Analysis — Spectrochim. Acta $\underline{22}$, 927 (1966) — V.B. Singh, R.N. Singh and I.S. Singh

m-Chloroaniline ($C_6H_4NH_2Cl$)

Raman — Monatsh. Chem. $\underline{65}$, 199 (1935) — K.W.F. Kohlrausch and A. Pongratz

Vibrational Analysis — J. Chem. Soc. 3670 (1959) — A.R. Katritzky and R.A. Jones

UV Absorption in Vapour — Indian J. Phys. $\underline{31}$, 577 (1957) — P.B.V. Harnath and K. Sreeramamurty

IR Spectrum — Indian J. Phys. $\underline{37}$, 275 (1963) — K.C. Medhi and G.S. Kastha

Vibrational Analysis — Spectrochim. Acta $\underline{22}$, 927 (1966) — V.B. Singh, R.N. Singh and I.S. Singh

p-Chloroaniline ($C_6H_4NH_2Cl$)

Raman — Monatsh. Chem. $\underline{72}$, 1268 (1939) — K.W.F. Kohlrausch and O. Paulsen

UV Absorption in Vapour — Indian J. Phys. $\underline{31}$, 577 (1957) — P.B.V. Harnath and K. Sreeramamurty

Vibrational Analysis — Spectrochim. Acta $\underline{12}$, 305 (1958) — C. Garrigou-Lagrange, J.M. Lebas and M.L. Josien

Vibrational Analysis — J. Chem. Soc. 3670 (1959) — A.R. Katritzky and R.N. Jones

Vibrational Analysis — Spectrochim. Acta $\underline{13}$, 225 (1959) — J.M. Lebas, C. Garri-gou-Lagrange and M.L. Josien

IR Spectrum — Indian J. Phys. $\underline{37}$, 275 (1963) — K.C. Medhi and G.S. Kastha

Vibrational Analysis — Spectrochim. Acta $\underline{22}$, 927 (1966) — V.B. Singh, R.N. Singh and I.S. Singh

Calculation — J. Mol. Struct. $\underline{14}$, 61 (1972)— L. Smetankine and J. Etchepare

o-Chloroanisole ($C_6H_4OCH_3Cl$)

Raman — Monatsh. Chem. $\underline{76}$, 22 (1946) — E. Herz

UV Absorption in Vapour — J. Sci. Ind. Res. (India) $\underline{14B}$, 36 (1955) — V. Suryana-rayana and V.R. Rao

UV Absorption in Vapour — Indian J. Phys. $\underline{30}$, 117 (1956) — V. Suryanarayana and V.R. Rao

UV Absorption in Different States — Indian J. Phys. $\underline{31}$, 135 (1957) — S.B. Banerjee

UV Absorption in Vapour — Indian J. Phys. $\underline{31}$, 444 (1957) — S.L.N.G. Krishnama-chari

Raman — Indian J. Phys. $\underline{34}$, 247 (1960) — K.K. Deb

Vibrational Analysis — Spectrochim. Acta $\underline{19}$, 877 (1963) — E.F. Mooney

Vibrational Analysis — Spectrochim. Acta $\underline{25A}$, 343 (1969) — N.L. Owen and H.E. Hester

m-Chloroanisole ($C_6H_4OCH_3Cl$)

UV Absorption in Vapour — Indian J. Phys. $\underline{31}$, 577 (1957) — P.B.V. Harnath

Vibrational Analysis — Spectrochim. Acta $\underline{19}$, 877 (1963) — E.F. Mooney

UV Absorption in Vapour — Indian J. Pure Appl. Phys. $\underline{5}$, 189 (1967) — A.N. Pathak

Vibrational Analysis — Spectrochim. Acta $\underline{25A}$, 343 (1969) — N.L. Owen and R.E. Hester

p-Chloroanisole ($C_6H_4OCH_3Cl$)

IR Spectrum — J. Phys. Radium $\underline{9}$, 26 (1938) — J. Lecomte

Raman — Monatsh. Chem. $\underline{72}$, 244 (1939) — O. Paulsen

Raman — Monatsh. Chem. $\underline{74}$, 160 (1941) — E. Herz

Raman — Monatsh. Chem. $\underline{76}$, 200 (1947) — E. Herz, K.W.F. Kohlrausch and R. Yogel

UV Absorption in Vapour — J. Sci. Ind. Res. (India) $\underline{14B}$, 128 (1955) — V. Suryanarayana and V.R. Rao

UV Absorption in Vapour — Indian J. Phys. $\underline{31}$, 619 (1957) — V. Suryanarayana and V.R. Rao

UV Absorption in Different States — Indian J. Phys. $\underline{31}$, 135 (1957) — S.B. Banerjee

Vibrational Analysis — Spectrochim. Acta
12, 305 (1958) — C. Garrigou-Lagrange,
J.M. Lebas and M.L. Josien
Vibrational Analysis — Spectrochim. Acta
13, 225 (1959) — J.M. Lebas, C.
Garrigou-Lagrange and M.L. Josien
Raman — Indian J. Phys. 34, 247 (1960) —
K.K. Deb
UV Absorption in Solution — Spectrochim.
Acta 17, 545 (1961) — E. Spinner
Vibrational Analysis — Spectrochim. Acta
19, 877 (1963) — E.F. Mooney
Vibrational Analysis — Spectrochim. Acta
19, 807 (1963) — J.H.S. Green, W.
Kynaston and H.A. Gebbie
Vibrational Analysis — Spectrochim. Acta
22, 1427 (1966) — J.N. Rai and K.N.
Upadhya
Vibrational Analysis — Spectrochim. Acta
23A, 1111 (1967) — M. Horak, E.R.
Lippincott and R.K. Khanna

1-Chloroanthraquinone ($C_{14}H_7O_2Cl_2$)

IR Spectrum — Curr. Sci. 36, 483 (1967) —
S. Nath Singh and R.S. Singh
UV Absorption in Vapour — Curr. Sci. 37,
191 (1968) — K.P.R. Nair and S. Nath
Singh
IR Spectrum — Indian J. Pure Appl. Phys.
6, 454 (1968) — S. Nath Singh and
R.S. Singh
UV Absorption in Vapour — Indian J. Pure
Appl. Phys. 8, 641 (1970) — S. Nath
Singh and R.S. Singh

2-Chloroanthraquinone ($C_{14}H_7O_2Cl_2$)

IR Spectrum — Indian J. Pure Appl. Phys.
6, 454 (1968) — S. Nath Singh and
R.S. Singh
IR Spectrum — Indian J. Pure Appl. Phys.
7, 133 (1969) — S. Nath Singh, G.D.
Baruah and K.P.R. Nair
Emission Spectrum in Vapour — Indian J.
Pure Appl. Phys. 7, 133 (1969) — S.
Nath Singh, G.D. Baruah and K.P.R.
Nair
UV Absorption in Vapour — Indian J. Pure
Appl. Phys. 8, 641 (1970) — S. Nath
Singh and R.S. Singh

o-Chlorobenzaldehyde (C_6H_4CHOCl)

Raman — Z. Phys. Chem. 38B, 119 (1937) —
K.W.F. Kohlrausch and L. Kahovec
IR Spectrum — J. Phys. Radium 9, 13
(1938) — J. Lecomte
Raman — Z. Phys. Chem. B38, 119 (1939) —
K.W.F. Kohlrausch
UV Absorption in Vapour — J. Sci. Ind.
Res. (India) 18B, 265 (1959) — J.C.
Patel
UV Absorption in Vapour — Curr. Sci. 29,
344 (1960) — M.R. Padhye and B.G.
Viladkar
Vibrational Analysis — J. Sci. Ind. Res.
(India) 19B, 45 (1960) — M.R. Padhye
and B.G. Viladkar
UV Absorption in Vapour — J. Sci. Ind.
Res. (India) 20B, 523 (1961) — I.A.
Rao
Raman — J. Sci. Ind. Res. (India) 20B, 523
(1961) — I.A. Rao
UV Absorption in Vapour — J. Sci. Ind.
Res. (India) 20B, 530 (1961) — M.R.
Padhye and B.G. Viladkar
Emission Spectrum in Vapour — Indian J.
Pure Appl. Phys. 3, 342 (1965) — N.L.
Singh and D.P. Juyal
Emission Spectrum in Vapour — Curr. Sci.
34, 175 (1965) — D.P. Juyal
UV Absorption in Vapour — Indian J. Pure
Appl. Phys. 4, 92 (1966) — M.R. Padhye
and B.G. Viladkar
UV Absorption in Solution — Spectrochim.
Acta 28A, 1969 (1972) — E.V. Donckt
and C. Vogels
Vibrational Analysis — Spectrochim. Acta
32A, 1265 (1976) — J.H.S. Green and
D.J. Harrison

m-Chlorobenzaldehyde (C_6H_4CHOCl)

Raman — Z. Phys. Chem. 38B, 119 (1937) —
K.W.F. Kohlrausch and L. Kahovec
IR Spectrum — J. Phys. Radium 9, 13
(1938) — J. Lecomte
IR Spectrum — Anal. Chem. 29, 334 (1957) —
S. Pinchas
UV Absorption in Vapour — J. Sci. Ind.
Res. (India) 18B, 265 (1959) — J.C.
Patel
Vibrational Analysis — J. Sci. Ind. Res.
(India) 19B, 45 (1960) — M.R. Padhye
and B.G. Viladkar
UV Absorption in Vapour — Curr. Sci. 29,
344 (1960) — M.R. Padhye and B.G.
Viladkar
UV Absorption in Vapour — J. Sci. Ind.
Res. (India) 20B, 530 (1961) — M.R.
Padhye and B.G. Viladkar
Emission Spectrum in Vapour — Curr. Sci.
34, 175 (1965) — D.P. Juyal

UV Absorption in Vapour — Indian J. Pure
  Appl. Phys. 4, 92 (1966) — M.R. Padhye
  and B.G. Viladkar
Vibrational Analysis — J. Chim. Phys. 63,
  552 (1966) — C. Garrigou-Lagrange, M.
  Chehata and J. Lascombe
UV Absorption in Solution — Spectrochim.
  Acta 28A, 1969 (1972) — E.V. Danckt
  and C. Vogels
Vibrational Analysis — Spectrochim. Acta
  32A, 1265 (1976) — J.H.S. Green and
  D.J. Harrison

p-Chlorobenzaldehyde (C$_6$H$_4$CHOCl)

Raman — Z. Phys. Chem. 22B, 169 (1933) —
  G.B. Bonino and R. Manzoni-Ansidei
Raman — Z. Phys. Chem. 38B, 119 (1937) —
  K.W.F. Kohlrausch and L. Kahovec
IR Spectrum — J. Phys. Radium 9, 13 (1938)—
  J. Lecomte
IR Spectrum — Anal. Chem. 29, 334 (1957) —
  S. Pinchas
UV Absorption in Vapour — J. Sci. Ind.
  Res. (India) 18B, 265 (1959) — J.C.
  Patel
UV Absorption in Vapour — Curr. Sci. 29,
  344 (1960) — M.R. Padhye and B.G.
  Viladkar
Vibrational Analysis — J. Sci. Ind. Res.
  (India) 19B, 45 (1960) — M.R. Padhye
  and B.G. Viladkar
UV Absorption in Vapour — J. Sci. Ind.
  Res. (India) 20B, 530 (1961) — M.R.
  Padhye and B.G. Viladkar
Emission Spectrum in Vapour — Curr. Sci.
  34, 175 (1965) — D.P. Juyal
UV Absorption in Vapour — Indian J. Pure
  Appl. Phys. 4, 92 (1966) — M.R. Padhye
  and B.G. Viladkar
UV Absorption in Solution — Spectrochim.
  Acta 28B, 1969 (1972) — E.V. Danckt
  and C. Vogels
Vibrational Analysis — Spectrochim. Acta
  32A, 1265 (1976) — J.H.S. Green and
  D.J. Harrison

Chlorobenzene (C$_6$H$_5$Cl)

IR Spectrum — J. Chem. Soc. 931 (1936) —
  C.R. Bailey, J.B. Hale, C.K. Ingold
  and W.J. Thompson
UV Absorption in Vapour — Proc. Phys.
  Maths. Soc. Jap. 22, 677 (1940) — K.
  Asagoe and Y. Ikemoto
UV Absorption in Vapour — J. Chem. Phys.
  9, 667 (1941) — H. Sponer and J.S.
  Smith-Kirby

UV Absorption in Vapour — J. Chem. Phys.
  9, 814 (1941) — H. Sponer and S.H.
  Wollman
Vibrational Analysis — Proc. Indian Acad.
  Sci. A15, 401 (1942) — C.S. Venkate-
  swaran and N.S. Pandya
Fluorescence — J. Opt. Soc. Amer. 40, 389
  (1950) — A.M. Bass and H. Sponer
Raman — Indian J. Phys. 24, 131 (1950) —
  A.K. Roy
UV Absorption in Vapour — J. Chem. Phys.
  22, 234 (1954) — H. Sponer
IR Spectrum — J. Chem. Soc. 1350 (1956) —
  D.H. Whiffen
Raman — Indian J. Phys. 31, 635 (1957) —
  G.S. Kastha
UV Absorption in Solution — Can. J. Chem.
  38, 1104 (1960) — W.F. Forbes
IR Spectrum — Spectrochim. Acta 19, 601
  (1963) — J.R. Scherer
IR Spectrum — Indian J. Phys. 38, 610
  (1964) — S.C. Sirkar, D.K. Mukharjee
  and P.K. Bishnui
Calculation — Spectrochim. Acta 21, 321
  (1965) — J.R. Scherer
IR Spectrum — Spectrochim. Acta 22, 501
  (1966) — W.R. McWhinnie
UV Absorption Spectrum — C.R. Acad. Sci.B
  265, 641 (1967) — A. Quemerais, M.
  Merlais and S. Robin
Vibrational Analysis — Spectrochim. Acta
  26A, 1925 (1970) — J.H.S. Green
  and D.J. Harrison
UV Absorption in Vapour (High Resolution)—
  Mol. Phys. 18, 101 (1970) — T. Cvitas
  and J.M. Hollas
UV Absorption in Vapour — Appl. Spectrosc.
  24, 292 (1970) — H.D. Bist, V.N. Sarin,
  A. Ojhe and Y.S. Jain
UV Absorption in Vapour — Chem. Phys. Lett.
  4, 15 (1970) — A.L. Verma and H.D.
  Bist
Raman — J. Mol. Struct. 5, 477 (1970) —
  N.T. McDevitt and W.G. Fateley
UV Absorption in Vapour — Spectrochim.
  Acta 26A, 841 (1970) — H.D. Bist,
  V.N. Sarin, A. Ojhe and Y.S. Jain
Raman (High Resolution) — J. Mol. Spectr-
  osc. 39, 73 (1971) — C.T. Meneely,
  C.Y. She and D.F. Edwards
UV Absorption in Vapour — J. Mol.
  Spectrosc. 47, 126 (1973) — Y.S. Jain
  and H.D. Bist
Calculation — J. Mol. Struct. 16, 365
  (1973) — R.T.C. Brownlee, D.G. Cameron,
  R.D. Topsom, A.R. Katritzky and A.J.
  Sparrow

Chlorobenzene-o-D (C_6H_4ClD)

Vibrational Analysis — J. Chim. Phys. <u>64</u>,
1473 (1967) — G. Lucazeau and J.M.
Lebas

Chlorobenzene-m-D (C_6H_4ClD)

Vibrational Analysis — J. Chim. Phys. <u>64</u>,
1473 (1967) — G. Lucazeau and J.M.
Lebas

Chlorobenzene-p-D (C_6H_4ClD)

Vibrational Analysis — J. Chim. Phys. <u>64</u>,
1473 (1967) — G. Lucazeau and J.M.
Lebas

Chlorobenzene-D_5 (C_6D_5Cl)

Vibrational Analysis — Spectrochim. Acta
<u>21</u>, 1495 (1965) — T.R. Nanney, R.T.
Bailey and E.R. Lippincott
UV Absorption in Vapour — Indian J. Pure
Appl. Phys. <u>9</u>, 383 (1971) — R.S.
Tripathi and B.R. Pandey

p-Chlorobenzenethiol (C_6H_4ClSH)

Vibrational Analysis — Spectrochim. Acta
<u>17</u>, 795 (1961) — R.A. Nyquist and
J.C. Evans
Vibrational Analysis — Spectrochim. Acta
<u>26A</u>, 1515 (1970) — J.H.S. Green, D.J.
Harrison, W. Kynaston and D.W. Scott

o-Chlorobenzoic acid ($C_6H_4CO_2HCl$)

Vibrational Analysis — Spectrochim. Acta
<u>33A</u>, 575 (1977) — J.H.S. Green

m-Chlorobenzoic acid ($C_6H_4CO_2HCl$)

Vibrational Analysis — Spectrochim. Acta
<u>33A</u>, 575 (1977) — J.H.S. Green

p-Chlorobenzoic acid ($C_6H_4CO_2HCl$)

Vibrational Analysis — Spectrochim. Acta
<u>33A</u>, 575 (1977) — J.H.S. Green

o-Chlorobenzonitrile (C_6H_4CNCl)

Raman — Sber. Akad. Wiss Wien <u>143</u>, 551
(1934) — K.W.F. Kohlrausch and A.
Pongratz
IR Spectrum — Indian J. Pure Appl. Phys.
<u>4</u>, 169 (1966) — S.M. Pandey and B.R.
Pandey
UV Absorption in Vapour — Indian J. Pure
Appl. Phys. <u>4</u>, 169 (1966) — S.M.
Pandey and B.R. Pandey
UV Absorption in Vapour — Indian J. Pure
Appl. Phys. <u>11</u>, 341 (1973) — T.S.
Varadarajan and S. Parthasarthy
Vibrational Analysis — Spectrochim. Acta
<u>32A</u>, 1279 (1976) — J.H.S. Green and
D.J. Harrison

m-Chlorobenzonitrile (C_6H_4CNCl)

UV Absorption in Vapour — Indian J. Pure
Appl. Phys. <u>11</u>, 341 (1973) — T.S.
Varadarajan and S. Parthasarthy
Vibrational Analysis — Spectrochim. Acta
<u>32A</u>, 1279 (1976) — J.H.S. Green and
D.J. Harrison

p-Chlorobenzonitrile (C_6H_4CNCl)

Raman — Sber. Akad. Wiss Wien <u>143</u>, 551
(1934) — K.W.F. Kohlrausch and A.
Pongratz
Raman — Sber. Akad. Wiss Wien <u>144</u>, 417
(1935) — K.W.F. Kohlrausch and G.P.
Ypsilanti
Raman — Monatsh. Chem. <u>65</u>, 199 (1935) —
K.W.F. Kohlrausch and A. Pongratz
Vibrational Analysis — Spectrochim. Acta
<u>12</u>, 305 (1958) — C. Garrigou-Lagrange,
J.M. Lebas and M.L. Josien
Vibrational Analysis — Spectrochim. Acta
<u>13</u>, 225 (1959) — J.M. Lebas, C.
Garrigou-Lagrange and M.L. Josien
Vibrational Analysis — Z. Elektrochem.
<u>64</u>, 1228 (1960) — S. Weckherlin and
W. Luttke
Vibrational Analysis — Spectrochim. Acta
<u>21</u>, 45 (1965) — H.W. Wilson and J.E.
Bloor
UV Absorption in Vapour — Indian J. Pure
Appl. Phys. <u>4</u>, 169 (1966) — S.M.
Pandey and B.R. Pandey
IR Spectrum — Indian J. Pure Appl. Phys.
<u>4</u>, 169 (1966) — S.M. Pandey and B.R.
Pandey
UV Absorption in Vapour — Indian J. Pure
Appl. Phys. <u>6</u>, 84 (1968) — B.R. Pandey

Vibrational Analysis — J. Chem. Soc. Perkin
II 1607 (1972) — P.N. Gates, D. Steele,
R.A.R. Pearce and K. Radcliffe
UV Absorption in Vapour — Indian J. Pure
Appl. Phys. $\underline{11}$, 341 (1973) — T.S.
Varadarajan and S. Parthasarthy
Vibrational Analysis — Spectrochim. Acta
$\underline{32A}$, 1279 (1976) — J.H.S. Green and
D.J. Harrison

2-Chloro-p-benzoquinone-D$_3$ (C$_6$D$_3$ClO$_2$)

Vibrational Analysis — Spectrochim. Acta
$\underline{34A}$, 453 (1978) — A. Girlands and C.
Pecile

o-Chlorobenzotrifluoride (C$_6$H$_4$CF$_3$Cl)

UV Absorption in Vapour — J. Chem. Phys.
$\underline{27}$, 740 (1957) — N.A. Narasimhan,
J.R. Nielsen and R. Theimer
UV Absorption in Vapour — J. Mol. Spectr-
osc. $\underline{4}$, 43 (1960) — I.A. Rao and V.R.
Rao
Vibrational Analysis — J. Mol. Spectrosc.
$\underline{6}$, 447 (1961) — I.A. Rao and V.R. Rao
UV Absorption in Vapour — Indian J. Pure
Appl. Phys. $\underline{6}$, 205 (1968) — P.D. Singh
Vibrational Analysis — Spectrochim. Acta
$\underline{33A}$, 837 (1977) — J.H.S. Green and
D.J. Harrison

p-Chlorobenzotrifluoride (C$_6$H$_4$CF$_3$Cl)

UV Absorption in Vapour — J. Chem. Phys.
$\underline{27}$, 740 (1957) — N.A. Narasimhan, J.R.
Nielsen and R. Theimer
UV Absorption in Vapour — J. Mol. Spectrosc.
$\underline{4}$, 43 (1960) — I.A. Rao and V.R. Rao
Vibrational Analysis — J. Mol. Spectrosc.
$\underline{6}$, 447 (1961) — I.A. Rao and V.R. Rao
UV Absorption in Vapour — Indian J. Pure
Appl. Phys. $\underline{6}$, 205 (1968) — P.D. Singh
Vibrational Analysis — Spectrochim. Acta
$\underline{33A}$, 837 (1977) — J.H.S. Green and
D.J. Harrison

m-Chlorobenzylbromide (C$_6$H$_4$CH$_2$BrCl)

Vibrational Analysis — Spectrochim. Acta
$\underline{29A}$, 813 (1973) — L. Verdonck, G.P.
Van der Kelen and Z. Eeckhaut

m-Chlorobenzylchloride (C$_6$H$_4$CH$_2$ClCl)

Vibrational Analysis — Spectrochim. Acta
$\underline{29A}$, 813 (1973) — L. Verdonck, G.P.
Van der Kelen and Z. Eeckhaut

p-Chlorobenzylchloride (C$_6$H$_4$CH$_2$ClCl)

Vibrational Analysis — Spectrochim. Acta
$\underline{28A}$, 55 (1972) — L. Verdonck and G.P.
Van der Kelen

1-Chloro-2,4-Dinitrobenzene (C$_6$H$_3$Cl(NO$_2$)$_2$)

Vibrational Analysis — Indian J. Phys. $\underline{35}$,
583 (1961) — K.C. Medhi
Vibrational Analysis — Spectrochim. Acta
$\underline{20}$, 1021 (1964) — E.F. Mooney

Chloroethynylbenzene (C$_6$H$_5$C:CCl)

Vibrational Analysis — Spectrochim. Acta
$\underline{26A}$, 919 (1970) — R.D. McLachelaun

o-Chloroethylbenzene (C$_6$H$_4$C$_2$H$_5$Cl)

Raman — Indian J. Phys. $\underline{29}$, 254 (1955) —
S.L.N.G. Krishnamachari and V.R. Rao
UV Absorption in Vapour — J. Sci. Ind.
Res. (India) $\underline{14B}$, 546 (1955) — U.D.
Gantajet and V.R. Rao
UV Absorption in Different States — Indian
J. Pure Appl. Phys. $\underline{7}$, 62 (1969) —
K.G. Srinivasacharya

m-Chloroethylbenzene (C$_6$H$_4$C$_2$H$_5$Cl)

UV Absorption in Vapour — J. Sci. Ind.
Res. (India) $\underline{15B}$, 549 (1956) — G.
Sitharmamurty and V.R. Rao
UV Absorption in Different States — Indian
J. Pure Appl. Phys. $\underline{7}$, 62 (1969) — K.G.
Srinivasacharya

o-Chlorofluorobenzene (C$_6$H$_4$ClF)

UV Absorption in Solution — Z. Phys. Chem.
$\underline{33}$, 311 (1936) — H. Conrad-Billroth
and G. Forster
UV Absorption in Vapour — Indian J. Phys.
$\underline{29}$, 603 (1955) — S.L.N.G. Krishnamachari
Vibrational Analysis — J. Chem. Phys. $\underline{24}$,
1232 (1956) — F.W. Harris, N.A. Naras-
imhan and J.R. Nielsen

Vibrational Analysis — J. Chem. Phys. <u>24</u>, 440 (1956) — N.A. Narasimhan and J.R. Nielsen

Vibrational Analysis — J. Chem. Phys. <u>24</u>, 433 (1956) — N.A. Narasimhan and J.R. Nielsen

Raman — Curr. Sci. <u>25</u>, 185 (1956) — S.L. N.G. Krishnamachari

UV Absorption in Vapour — Indian J. Phys. <u>31</u>, 447 (1957) — S.L.N.G. Krishnamachari

UV Absorption Spectrum — Spectrochim. Acta <u>19</u>, 683 (1963) — G. Varsanyi, S. Holly and T. Farago

Calculation — Curr. Sci. <u>35</u>, 331 (1966) — N.A. Narasimhan and C.V.S.R. Rao

Normal Coordinate Analysis — J. Mol. Spectrosc. <u>28</u>, 44 (1968) — N.A. Narasimhan and C.V.S.R. Rao

UV Absorption in Different States — Indian J. Pure Appl. Phys. <u>7</u>, 62 (1969) — K.G. Srinivasacharya

Normal Coordinate Analysis — J. Mol. Spectrosc. <u>30</u>, 192 (1969) — N.A. Narasimhan and C.V.S.R. Rao

Vibrational Analysis — Spectrochim. Acta <u>26A</u>, 1913 (1970) — J.H.S. Green

o-Chlorofluorobenzene-D_4 (C_6D_4ClF)

Vibrational Analysis — J. Mol. Spectrosc. <u>48</u>, 185 (1973) — N.D. Patel, V.B. Kartha and N.A. Narasimhan

m-Chlorofluorobenzene (C_6H_4ClF)

UV Absorption in Solution — Z. Phys. Chem. <u>33</u>, 311 (1936) — H. Conrad-Billroth and G. Forster

Raman — Bull. Soc. Rou. Phys. <u>43</u>, 43 (1942) — H. Tintea

UV Absorption in Vapour — Indian J. Phys. <u>30</u>, 151 (1956) — S.L.N.G. Krishnamachari

Vibrational Analysis — J. Chem. Phys. <u>24</u>, 1232 (1956) — F.W. Harris, N.A. Narasimhan and J.R. Nielsen

Vibrational Analysis — J. Chem. Phys. <u>24</u>, 433 (1956) — N.A. Narasimhan and J.R. Nielsen

UV Absorption in Vapour — Indian J. Phys. <u>31</u>, 447 (1957) — S.L.N.G. Krishnamachari

Vibrational Analysis — J. Chim. Phys. <u>63</u>, 552 (1966) — C. Garrigou-Lagrange, M. Chehata and J. Lascombe

Calculation — Curr. Sci. <u>35</u>, 331 (1966) — N.A. Narasimhan and C.V.S.R. Rao

UV Emission in Vapour — J. Mol. Spectrosc. <u>22</u>, 148 (1967) — J.N. Rai and K.N. Upadhya

Calculation — J. Mol. Spectrosc. <u>28</u>, 44 (1968) — N.A. Narasimhan and C.V.S.R. Rao

UV Absorption in Different States — Indian J. Pure Appl. Phys. <u>7</u>, 62 (1969) — K.G. Srinivasacharya

Normal Coordinate Analysis — J. Mol. Spectrosc. <u>30</u>, 192 (1969) — N.A. Narasimhan and C.V.S.R. Rao

Vibrational Analysis — Spectrochim. Acta <u>26A</u>, 1523 (1970) — J.H.S. Green

UV Absorption in Vapour (High Resolution) — J. Mol. Spectrosc. <u>39</u>, 345 (1971) — C.Y. Wu and J.R. Lombardi

m-Chlorofluorobenzene-D_4 (C_6D_4ClF)

Vibrational Analysis — J. Mol. Spectrosc. <u>48</u>, 185 (1973) — N.D. Patel, V.B. Kartha and N.A. Narasimhan

p-Chlorofluorobenzene (C_6H_4ClF)

Raman — Monatsh. Chem. <u>76</u>, 200 (1947) — K.W.F. Kohlrausch

Vibrational Analysis — J. Chem. Phys. <u>24</u>, 420 (1956) — N.A. Narasimhan, M.Z. El Sabban and J.R. Nielsen

UV Absorption in Vapour — Indian J. Phys. <u>30</u>, 319 (1956) — S.L.N.G. Krishnamachari

Vibrational Analysis — J. Chem. Phys. <u>24</u>, 433 (1956) — N.A. Narasimhan and J.R. Nielsen

UV Absorption in Vapour — Indian J. Phys. <u>31</u>, 447 (1957) — S.L.N.G. Krishnamachari

Vibrational Analysis — Spectrochim. Acta <u>12</u>, 57 (1958) — A. Stojiikovic and D.H. Whiffen

IR Spectrum — J. Sci. Ind. Res. (India) — <u>18B</u>, 504 (1959) — M.R. Padhye and B.G. Viladkar

Calculation — Curr. Sci. <u>35</u>, 331 (1966) — N.A. Narasimhan and C.V.S.R. Rao

Vibrational Analysis — Proc. Roy. Soc. <u>A298</u>, 51 (1967) — P.R. Griffiths and H.W. Thompson

Vibrational Analysis — J. Chim. Phys. <u>64</u>, 1450 (1967) — P. Pommez, M. Lafaix, P. Delorme and V. Lorenzelli

Normal Coordinate Analysis – J. Mol.
    Spectrosc. 28, 144 (1968) – N.A.
    Narasimhan and C.V.S.R. Rao
UV Absorption in Vapour (High Resolution) –
    J. Mol. Struct. 4, 459 (1969) – S.N.
    Thakur and K.N. Upadhya
UV Absorption in Different States – Indian
    J. Pure Appl. Phys. 7, 62 (1969) – K.G.
    Srinivasacharya
Normal Coordinate Analysis – J. Mol.
    Spectrosc. 30, 192 (1969) – N.A.
    Narasimhan and C.V.S.R. Rao
Vibrational Analysis – Spectrochim. Acta
    26A, 1503 (1970) – J.H.S. Green
UV Absorption in Vapour (High Resolution) –
    Mol. Phys. 18, 261 (1970) – T. Cvitas
    and J.M. Hollas

p-Chlorofluorobenzene-D$_4$ (C$_6$D$_4$ClF)

Vibrational Analysis – J. Mol. Spectrosc.
    48, 185 (1973) – N.D. Patel, V.B.
    Kartha and N.A. Narasimhan

2-Chloro-4-Fluorotoluene (C$_6$H$_3$ClFCH$_3$)

UV Absorption in Vapour – Indian J. Pure
    Appl. Phys. 10, 762 (1972) – V.N. Verma
Vibrational Analysis – Spectrochim. Acta
    27A, 807 (1971) – J.H.S. Green, D.J.
    Harrison and W. Kynaston
Vibrational Analysis – Indian J. Phys.
    49, 851 (1975) – V.N. Verma
IR Spectrum – Curr. Sci. 44, 218 (1975) –
    D.R. Singh, U.S. Tripathi, R.C.
    Maheshwari and D. Sharma

2-Chloro-6-Fluorotoluene (C$_6$H$_3$ClFCH$_3$)

Vibrational Analysis – Spectrochim. Acta
    27A, 793 (1971) – J.H.S. Green, D.J.
    Harrison and W. Kynaston
UV Absorption in Vapour – Indian J. Pure
    Appl. Phys. 11, 866 (1973) – G.N.R.
    Tripathi, R.M. Verma, B.N. Tewari and
    S. Ram

4-Chloro-2-Fluorotoluene (C$_6$H$_3$ClFCH$_3$)

UV Absorption in Vapour – Indian J. Pure
    Appl. Phys. 9, 137 (1971) – V.N. Verma
Vibrational Analysis – Spectrochim. Acta
    27A, 807 (1971) – J.H.S. Green, D.J.
    Harrison and W. Kynaston
IR Spectrum – Indian J. Phys. 48, 854
    (1974) – G. Joshi and K.K. Sharma

5-Chloro-2-Fluorotoluene(C$_6$H$_3$ClFCH$_3$)

Vibrational Analysis – Spectrochim. Acta
    27A, 807 (1971) – J.H.S. Green, D.J.
    Harrison and W. Kynaston
UV Absorption in Vapour – Curr. Sci. 44,
    218 (1975) – D.R. Singh, U.S. Tripathi,
    R.C. Maheshwari and D. Sharma
IR Spectrum – Curr. Sci. 44, 218 (1975) –
    D.R. Singh, U.S. Tripathi, R.C.
    Maheshwari and D. Sharma

6-Chloro-2-Fluorotoluene (C$_6$H$_3$ClFCH$_3$)

Vibrational Analysis – Spectrochim. Acta
    27A, 793 (1971) – J.H.S. Green, D.J.
    Harrison and W. Kynaston

6-Chloro-3-Fluorotoluene (C$_6$H$_3$ClFCH$_3$)

Vibrational Analysis – Spectrochim. Acta
    27A, 807 (1971) – J.H.S. Green, D.J.
    Harrison and W. Kynaston
UV Absorption in Vapour – Curr. Sci. 44,
    218 (1975) – D.R. Singh, U.S. Tripathi,
    R.C. Maheshwari and D. Sharma

2-Chloro-6-Hydroxypyridine(C$_5$H$_3$NClOH)

IR Spectrum – Indian J. Pure Appl. Phys.
    13, 416 (1975) – R.S. Tripathi and B.R.
    Pandey

o-Chloroiodobenzene (C$_6$H$_4$ClI)

Raman – Monatsh. Chem. 65, 199 (1935) –
    K.W.F. Kohlrausch and A. Pongratz
Vibrational Analysis – Spectrochim. Acta
    26A, 1913 (1970) – J.H.S. Green
IR Spectrum – Spectrochim. Acta 29A,
    1555 (1973) – J.M. Briody

m-Chloroiodobenzene (C$_6$H$_4$ClI)

Vibrational Analysis – J. Chim. Phys. 63,
    552 (1966) – C. Garrigou-Lagrange, M.
    Chehata and J. Lascombe
Vibrational Analysis – Spectrochim. Acta
    26A, 1523 (1970) – J.H.S. Green

p-Chloroiodobenzene (C_6H_4ClI)

Vibrational Analysis — Spectrochim. Acta 12, 57 (1958) — A. Stojilkovic and D.H. Whiffen

Vibrational Analysis — Spectrochim. Acta 12, 305 (1958) — C. Garrigou-Lagrange, J.M. Lebas and M.L. Josien

Vibrational Analysis — Spectrochim. Acta 13, 225 (1959) — J.M. Lebas, C. Garrigou-Lagrange and M.L. Josien

Vibrational Analysis — J. Chim. Phys. 64, 1450 (1967) — P. Pommez, M. Lafaix, P. Delorme and V. Lorenzelli

Vibrational Analysis — Proc. Roy. Soc. A298, 51 (1967) — P.R. Griffiths and H.W. Thompson

UV Absorption in Different States — Indian J. Pure Appl. Phys. 7, 62 (1969) — K.G. Srinivasacharya

Vibrational Analysis — Spectrochim. Acta 26A, 1503 (1970) — J.H.S. Green

IR Spectrum — Spectrochim. Acta 29A, 1555 (1973) — J.M. Briody

p-Chloro-N-Methylaniline ($C_6H_4ClNH:CH_3$)

IR Spectrum — J. Chem. Soc. 843 (1957) — D. Hadzi and M. Skrbljak

α-Chloronaphthalene ($C_{10}H_7Cl$)

UV Absorption in Vapour — Proc. Roy. Soc. A111, 355 (1926) — H.G. De Lazlo

UV Absorption in Different States — Indian J. Phys. 28, 21 (1954) — A.R. Deb

Phosphorescence — J. Chem. Soc. 3160 (1954) — J. Ferguson, T. Iredale and J.A. Taylor

UV Absorption in Solution — J. Chem. Soc. 304 (1954) — J. Ferguson

Vibrational Analysis — Indian J. Phys. 36, 557 (1962) — K.K. Deb

UV Absorption in Vapour — Indian J. Pure Appl. Phys. 4, 263 (1966) — R.D. Singh and R.S. Singh

IR Spectrum — Indian J. Pure Appl. Phys. 8, 348 (1970) — R.D. Singh and R.S. Singh

UV Absorption in Vapour — Indian J. Pure Appl. Phys. 14, 583 (1976) — R.D. Singh and S.N. Singh

Vibrational Analysis — Spectrochim. Acta 34A, 985 (1978) — S.N. Singh, H.S. Bhatti and R.D. Singh

β-Chloronaphthalene ($C_{10}H_7Cl$)

UV Absorption in Solution — Proc. Roy. Soc. A111, 355 (1926) — H.G. De Lazlo

Raman — Z. Elektrochem. 52, 210 (1948) — V.H. Luther

IR Spectrum — J. Chem. Soc. 3645 (1954) — J. Ferguson and R.L. Werner

Phosphorescence — J. Chem. Soc. 3160 (1954) — J. Ferguson, T. Iredale and J.A. Taylor

UV Absorption in Solution — J. Chem. Soc. 304 (1959) — J. Ferguson

UV Absorption in Different States — Indian J. Phys. 36, 302 (1962) — T.N. Misra

Vibrational Analysis — Indian J. Pure Appl. Phys. 10, 885 (1972) — O.P. Sharma and R.D. Singh

UV Absorption in Vapour — Indian J. Pure Appl. Phys. 14, 583 (1976) — R.D. Singh and S.N. Singh

2-Chloro-4-Nitroaniline ($C_6H_3ClNO_2NH_2$)

Vibrational Analysis — Indian J. Pure Appl. Phys. 8, 682 (1970) — V.N. Verma and K.P.R. Nair

Vibrational Analysis — Spectrosc. Lett. 6, 23 (1973) — V.N. Verma

4-Chloro-2-nitroaniline ($C_6H_3ClNO_2NH_2$)

Vibrational Analysis — Indian J. Pure Appl. Phys. 8, 682 (1970) — V.N. Verma and K.P.R. Nair

Vibrational Analysis — Spectrosc. Lett. 6, 23 (1973) — V.N. Verma

o-Chloronitrobenzene ($C_6H_4ClNO_2$)

Raman — Z. Phys. Chem. 52B, 315 (1942) — H. Witteck

Raman — Z. Elektrochem. 62, 544 (1958) — J. Behringer

Vibrational Analysis — Spectrochim. Acta 20, 1021 (1964) — E.F. Mooney

Vibrational Analysis — Spectrochim. Acta 26A, 1925 (1970) — J.H.S. Green and D.J. Harrison

m-Chloronitrobenzene ($C_6H_4ClNO_2$)

Raman — Z. Elektrochem. 62, 544 (1958) — J. Behringer

Vibrational Analysis — Spectrochim. Acta
   20, 1021 (1964) — E.F. Mooney
Vibrational Analysis — J. Chim. Phys. 63,
   552 (1966) — C. Garrigou-Lagrange, M.
   Chehata and J. Lascombe
Vibrational Analysis — Spectrochim. Acta
   26A, 1925 (1970) — J.H.S. Green and
   D.J. Harrison

p-Chloronitrobenzene ($C_6H_4ClNO_2$)

Raman — Z. Phys. Chem. 52B, 315 (1942) —
   H. Witteck
IR Spectrum — J. Amer. Chem. Soc. 77, 6341
   (1955) — J.F. Brown, Jr
IR Spectrum — J. Amer. Chem. Soc. 78, 4225
   (1956) — R.D. Kross and V.A. Fassel
Vibrational Analysis — Spectrochim. Acta
   12, 305 (1958) — C. Garrigou-Lagrange,
   J.M. Lebas and M.L. Josien
Vibrational Analysis — Spectrochim. Acta
   20, 1021 (1964) — E.F. Mooney
Vibrational Analysis — Indian J. Phys. 39,
   390 (1965) — K.C. Medhi
Vibrational Analysis — Proc. Roy. Soc.
   A298, 51 (1967) — P.R. Griffiths and
   H.W. Thompson
Vibrational Analysis — Spectrochim. Acta
   26A, 1925 (1970) — J.H.S. Green and
   D.J. Harrison
Calculation — J. Mol. Struct. 14, 61
   (1972) — L. Smetankine and J. Etchepare

p-Chloro-m-Nitrotoluene ($C_6H_3ClNO_2CH_3$)

IR Spectrum — Indian J. Pure Appl. Phys.
   16, 37 (1978) — M. Rangacharyulu and
   D. Premswarup

Chloropentafluorobenzene (C_6F_5Cl)

Vibrational Analysis — Spectrochim. Acta
   19, 1955 (1963) — D.A. Long and D.
   Steele
Vibrational Analysis — Spectrochim. Acta
   22, 695 (1966) — I.J. Hyams, E.R.
   Lippincott and R.T. Bailey
Vibrational Analysis — Spectrochim. Acta
   31A, 1839 (1975) — S.G. Frankiss and
   D.J. Harrison

o-Chlorophenetole ($C_6H_4OC_2H_5Cl$)

IR Spectrum — J. Amer. Chem. Soc. 58, 94
   (1936) — L. Pauling

Vibrational Analysis — Spectrochim. Acta
   19, 877 (1963) — E.F. Mooney
UV Absorption in Vapour — Indian J. Pure
   Appl. Phys. 3, 263 (1965) — K.V.K. Row

m-Chlorophenetole ($C_6H_4ClOC_2H_5$)

Vibrational Analysis — Spectrochim. Acta
   19, 877 (1963) — E.F. Mooney

p-Chlorophenetole ($C_6H_4ClOC_2H_5$)

Vibrational Analysis — Spectrochim. Acta
   19, 877 (1963) — E.F. Mooney
UV Absorption in Vapour — Indian J. Pure
   Appl. Phys. 3, 263 (1965) — K.V.K. Row

o-Chlorophenol (C_6H_4OHCl)

IR Spectrum — J. Phys. (Paris) 6, 281
   (1935) — J. Errera and P. Mollet
Calculation — J. Amer. Chem. Soc. 58, 44
   (1936) — L. Pauling
Raman — Dokl. Akad. Nauk. SSSR 47, 100
   (1945) — M.I. Batuev
Raman — Monatsh. Chem. 76, 1 (1946) — E.
   Herz
Raman — Monatsh. Chem. 76, 200 (1947) —
   E. Herz, K.W.F. Kohlrausch and R.
   Vogel
Raman — Monatsh. Chem. 76, 215 (1947) —
   K.W.F. Kohlrausch
UV Absorption in Vapour — Proc. Nat. Inst.
   Sci. (India) 17, 349 (1951) — C.
   Ramasastry
Raman — Indian J. Phys. 28, 85 (1954) —
   D.C. Biswas
Vibrational Analysis — Indian J. Phys.
   29, 264 (1955) — D.C. Biswas
Raman — J. Chem. Soc. 4314 (1956) — A.
   Buraway and P.R. Thompson
IR Spectrum — Indian J. Phys. 32, 345
   (1958) — S.C. Sirkar, A.R. Deb and
   S.B. Banerjee
Raman — Indian J. Phys. 32, 192 (1958) —
   D.K. Mukharjee
IR Spectrum — Appl. Spectrosc. 16, 32
   (1962) — R.J. Jackobsen and E.J.
   Brewer
IR Spectrum — Indian J. Pure Appl. Phys.
   4, 214 (1966) — K. Chandra
Vibrational Analysis — Spectrochim. Acta
   27A, 2199 (1971) — J.H.S. Green, D.J.
   Harrison and W. Kynaston

m-Chlorophenol (C_6H_4OHCl)

Raman — Monatsh. Chem. <u>65</u> 199 (1935) — K.W.F. Kohlrausch and A. Pongratz

Raman — Monatsh. Chem. <u>76</u>, 1 (1946) — E. Herz

Raman — Monatsh. Chem. <u>76</u>, 200 (1947) — E. Herz, K.W.F. Kohlrausch and R. Vogel

Raman — Monatsh. Chem. <u>76</u>, 215 (1947) — K.W.F. Kohlrausch

Vibrational Analysis — Indian J. Phys. <u>29</u>, 264 (1955) — D.C. Biswas

UV Absorption in Different States — Indian J. Phys. <u>35</u>, 203 (1961) — T.N. Misra and S.B. Banerjee

IR Spectrum — Appl. Spectrosc. <u>16</u>, 32 (1962) — R.J. Jackobsen and E.J. Brewer

UV Emission in Vapour — Indian J. Pure Appl. Phys. <u>3</u>, 70 (1965) — P.K. Verma

IR Spectrum — Indian J. Pure Appl. Phys. <u>4</u>, 214 (1966) — K. Chandra

Vibrational Analysis — J. Chim. Phys. <u>63</u>, 552 (1966) — C. Garrigou-Lagrange, M. Chehata and J. Lascombe

Vibrational Analysis — Spectrochim. Acta <u>27A</u>, 2199 (1971) — J.H.S. Green, D.J. Harrison and W. Kynaston

p-Chlorophenol (C_6H_4OHCl)

UV Absorption in Solution — J. Amer. Chem. Soc. 1088 (1913) — J.E. Purvis and A. McCleand

UV Absorption in Vapour — J. Amer. Chem. Soc. 1088 (1913) — J.E. Purvis and A. McCleand

Raman — Monatsh. Chem. <u>76</u>, 1 (1946) — E. Herz

UV Absorption in Solution — J. Amer. Chem. Soc. <u>69</u>, 2114 (1947) — L. Doubl and J.M. Vandenbelt

Raman — Monatsh. Chem. <u>76</u>, 215 (1947) — K.W.F. Kohlrausch

Raman — Monatsh. Chem. <u>76</u>, 200 (1947) — E. Herz, K.W.F. Kohlrausch and R. Vogel

UV Absorption in Vapour — Curr. Sci. <u>20</u>, 65 (1951) — C. Ramasastry

Raman — Indian J. Phys. <u>28</u>, 85 (1954) — D.C. Biswas

Vibrational Analysis — Indian J. Phys. <u>29</u>, 264 (1955) — D.C. Biswas

Vibrational Analysis — Spectrochim. Acta <u>12</u>, 305 (1958) — C. Garrigou-Lagrange, J.M. Lebas and M.L. Josien

Vibrational Analysis — Spectrochim. Acta <u>13</u>, 225 (1959) — J.M. Lebas, C. Garrigou-Lagrange and M.L. Josien

IR Spectrum — Appl. Spectrosc. <u>16</u>, 32 (1962) — R.J. Jackobsen and E.J. Brewer

IR Spectrum — Indian J. Pure Appl. Phys. <u>4</u>, 214 (1966) — K. Chandra

UV Emission in Vapour — Indian J. Pure Appl. Phys. <u>6</u>, 344 (1968) — K.N. Upadhya

UV Absorption in Vapour — Indian J. Pure Appl. Phys. <u>6</u>, 344 (1968) — K.N. Upadhya

Vibrational Analysis — Spectrochim. Acta <u>27A</u>, 2199 (1971) — J.H.S. Green, D.J. Harrison and W. Kynaston

o-Chloropyrazine ($C_4H_3N_2Cl$)

UV Absorption in Solution — J. Chem. Phys. <u>19</u>, 711 (1951) — F. Halverson and R.C. Hirt

UV Absorption in Solution — Spectrochim. Acta <u>12</u>, 114 (1958) — R.C. Hirt

UV Absorption in Vapour — Indian J. Phys. <u>46</u>, 306 (1972) — P.C. Upadhya, D.K. Rai, K.N. Upadhya and P.C. Mishra

α-Chloropyridine (C_5H_4NCl)

Vibrational Analysis — Spectrochim. Acta <u>19</u>, 549 (1963) — J.H.S. Green, W. Kynaston and H.M. Paisley

β-Chloropyridine (C_5H_4NCl)

Vibrational Analysis — Spectrochim. Acta <u>19</u>, 549 (1963) — J.H.S. Green, W. Kynaston and H.M. Paisley

UV Absorption in Vapour — Indian J. Pure Appl. Phys. <u>5</u>, 288 (1967) — B.R. Pandey

γ-Chloropyridine (C_5H_4NCl)

Vibrational Analysis — Spectrochim. Acta <u>19</u>, 549 (1963) — J.H.S. Green, W. Kynaston and H.M. Paisley

UV Absorption in Vapour — Indian J. Phys. <u>42</u>, 269 (1968) — B.R. Pandey

2-Chloropyrimidine ($C_4H_3N_2Cl$)

IR Spectrum — J. Chem. Soc. 168 (1952) — L.N. Short and H.W. Thompson

IR Spectrum — Spectrochim. Acta <u>26A</u>, 1243 (1970) — A.J. Lafaix and J.M. Lebas

Vibrational Analysis — Spectrochim. Acta
30A, 1801 (1974) — Y.A. Sarma
UV Absorption in Vapour — Indian J. Pure
Appl. Phys. 14, 689 (1976) — M.A.
Shashidhar and K.S. Rao
Vibrational Analysis — Spectrochim. Acta
33A, 189 (1977) — E. Allenstein, P.
Kiemle, J. Weidlein and W. Podszun

α-Chloroquinoline (C_9H_6NCl)

IR Spectrum — Indian J. Pure Appl. Phys.
8, 762 (1970) — M.A. Shashidhar
UV Absorption in Vapour — Indian J. Pure
Appl. Phys. 8, 178 (1970) — M.A.
Shashidhar and K.S. Rao

8-Chloroquinoline (C_9H_6NCl)

IR Spectrum — Indian J. Pure Appl. Phys.
8, 762 (1970) — M.A. Shashidhar
UV Absorption in Vapour — Indian J. Pure
Appl. Phys. 8, 178 (1970) — M.A.
Shashidhar and K.S. Rao

o-Chlorostyrene ($C_6H_4CH:CH_2Cl$)

Vibrational Analysis — Appl. Spectrosc. 22,
650 (1968) — W.G. Fateley, G.L. Garlson
and F.E. Dickson
Raman — Indian J. Pure Appl. Phys. 8, 146
(1970) — R.M.P. Jaiswal
IR Spectrum — Indian J. Pure Appl. Phys.
10, 55 (1972) — G.N.R. Tripathi

m-Chlorostyrene ($C_6H_4CH:CH_2Cl$)

Vibrational Analysis — Appl. Spectrosc.
22, 650 (1968) — W.G. Fateley, G.L.
Garlson and F.E. Dickson
Raman — Indian J. Pure Appl. Phys. 8,
146 (1970) — B. Singh and R.M.P.
Jaiswal
IR Spectrum — Indian J. Pure Appl. Phys.
10, 55 (1972) — G.N.R. Tripathi
Raman — J. Mol. Struct. 37, 85 (1977) —
L.A. Carreira and T.G. Towns

p-Chlorostyrene ($C_6H_4CH:CH_2Cl$)

Raman — Indian J. Pure Appl. Phys. 8,
146 (1970) — B. Singh and R.M.P.
Jaiswal
UV Absorption in Vapour — Indian J. Pure
Appl. Phys. 10, 239 (1972) — G.N.R.
Tripathi

IR Spectrum — Indian J. Pure Appl. Phys.
10, 55 (1972) — G.N.R. Tripathi
Raman — J. Mol. Struct. 37, 85 (1977) —
L.A. Carreira and T.G. Towns

p-Chlorothiophenol (C_6H_4SHCl)

Vibrational Analysis — Spectrochim. Acta
12, 305 (1958) — C. Garrigou-Lagrange,
J.M. Lebas and M.L. Josien
Vibrational Analysis — Spectrochim. Acta
17, 503 (1961) — R.A. Nyquist and J.C.
Evans
Vibrational Analysis — Spectrochim. Acta
26A, 1515 (1970) — J.H.S. Green, D.J.
Harrison and W. Kynaston
Vibrational Analysis — Spectrochim. Acta
26A, 847 (1970) — R.A. Nyquist
UV Absorption in Vapour — Indian J. Phys.
49, 703 (1975) — P.K. Mallik

o-Chlorotoluene ($C_6H_4CH_3Cl$)

UV Absorption in Solution — Z. Phys. Chem.
B13, 201 (1931) — K.L. Wolf and W.
Harold
IR Spectrum — J. Phys. Radium 9, 13
(1938) — J. Lecomte
Raman — Monatsh. Chem. 74, 16 (1943) —
E. Herz
Raman — Monatsh. Chem. 76, 13 (1947) — E.
Herz
UV Absorption in Different States — Indian
J. Phys. 26, 445 (1952) — H.N. Swamy
Vibrational Analysis — Indian J. Phys.
27, 447 (1953) — S.B. Sanyal
UV Absorption in Vapour — Curr. Sci. 22,
41 (1953) — G. Vishwanath
IR Spectrum — Indian J. Phys. 29, 257
(1955) — D.C. Biswas
UV Absorption in Solution — Indian J.
Phys. 30, 276 (1956) — S.B. Roy
Fluorescence — Indian J. Phys. 30, 255
(1956) — D.C. Biswas
Fluorescence — Indian J. Phys. 30, 407
(1956) — D.C. Biswas
Vibrational Analysis — J. Chem. Soc. 3670
(1959) — A.R. Katritzky and P. Simons
UV Absorption in Vapour — Indian J. Phys.
35, 143 (1961) — J.K. Roy
Vibrational Analysis — Spectrochim. Acta
20, 1343 (1964) — E.F. Mooney
IR Spectrum — Indian J. Phys. 39, 537
(1965) — D.K. Mukharjee, P.K. Bishnui
and S.C. Sirkar
Vibrational Analysis — J. Chim. Phys. 62,
347 (1965) — M. Brigodiot and J.M.
Lebas

IR Spectrum — Indian J. Pure Appl. Phys. 6, 488 (1968) — S.N. Singh and N.L. Singh

Vibrational Analysis — J. Sci. Res. BHU (India) 20(2), 25 (1969) — R. Amni Amma, K.P.R. Nair and D.K. Rai

UV Absorption in Vapour — Indian J. Pure Appl. Phys. 7, 768 (1969) — S. Nath Singh, R. Amni Amma and S.N. Singh

UV Emission in Vapour — Indian J. Pure Appl. Phys. 7, 768 (1969) — S. Nath Singh, R. Amni Amma and S.N. Singh

Vibrational Analysis — Spectrochim. Acta 26A, 1913 (1970) — J.H.S. Green

Raman — J. Mol. Struct. 5, 477 (1970) — N.T. McDevitt and W.G. Fateley

Calculation — J. Mol. Struct. 8, 319 (1971)— P.C. Mishra, S.N. Thakur and D.K. Rai

m-Chlorotoluene $(C_6H_4CH_3Cl)$

UV Absorption in Solution — Z. Phys. Chem. B13, 201 (1931) — K.L. Wolf and W. Harold

IR Spectrum — J. Phys. Radium 9, 13 (1938) — J. Lecomte

UV Absorption in Vapour — Bull. Sec. Sci. Akad. Rou. 22, 16 (1939) — H. Tintea

Raman — Monatsh. Chem. 74, 16 (1943) — E. Herz

Raman — Monatsh. Chem. 76, 13 (1947) — E. Herz

UV Absorption in Vapour — Indian J. Phys. 26, 263 (1952) — G. Vishwanath

UV Absorption in Different States — Indian J. Phys. 26, 445 (1952) — H.N. Swamy

UV Absorption in Vapour — Curr. Sci. 22, 41 (1953) — G. Vishwanath

Vibrational Analysis — Indian J. Phys. 29, 257 (1955) — D.C. Biswas

Fluorescence — Indian J. Phys. 29, 257 (1955) — D.C. Biswas

Fluorescence — Indian J. Phys. 30, 565 (1956) — D.C. Biswas

UV Absorption in Solution — Indian J. Phys. 30, 276 (1956) — S.B. Roy

Vibrational Analysis — J. Chem. Soc. 2058 (1959) — A.R. Katritzky and P. Simons

UV Absorption in Vapour — Indian J. Phys. 35, 628 (1961) — J.K. Roy

Vibrational Analysis — Spectrochim. Acta 20, 1343 (1964) — E.F. Mooney

IR Spectrum — Indian J. Phys. 39, 537 (1965) — D.K. Mukharjee, P.K. Bishnui and S.C. Sirkar

Raman — Indian J. Phys. 40, 415 (1966) — S.K. Nandy

Vibrational Analysis — J. Chim. Phys. 63, 552 (1966) — C. Garrigou-Lagrange, M. Chehata and J. Lascombe

Vibrational Analysis — Indian J. Pure Appl. Phys. 6, 488 (1968) — S.N. Singh and N.L. Singh

Vibrational Analysis — Spectrochim. Acta 26A, 1523 (1970) — J.H.S. Green

Vibrational Analysis — J. Sci. Res. BHU (India) 20(2), 25 (1970) — R. Amni Amma, K.P.R. Nair and D.K. Rai

Raman — J. Mol. Struct. 5, 477 (1970) — N.T. McDevitt and W.G. Fateley

UV Absorption in Vapour — Indian J. Pure Appl. Phys. 9, 330 (1971) — R. Amni Amma, S.N. Singh and K.P.R. Nair

Calculation — J. Mol. Struct. 8, 319 (1971) — P.C. Mishra, S.N. Thakur and D.K. Rai

Vibrational Analysis — Spectrochim. Acta 29A, 813 (1973) — L. Verdonck, G.P. Van der Kelen and Z. Eeckhaut

p-Chlorotoluene $(C_6H_4CH_3Cl)$

UV Absorption in Solution — Z. Phys. Chem. B13, 201 (1931) — K.L. Wolf and W. Harold

UV Absorption in Solution — Bull. Sec. Sci. Akad. Rou. 21 219 (1939) — H. Tintea

UV Absorption in Vapour — Indian J. Phys. 26, 263 (1952) — G. Vishwanath

UV Absorption in Different States — Indian J. Phys. 26, 445 (1952) — H.N. Swamy

Vibrational Analysis — Indian J. Phys. 27, 447 (1953) — S.B. Sanyal

Fluorescence — Indian J. Phys. 30, 143 (1956) — D.C. Biswas

UV Absorption in Solution — Indian J. Phys. 30, 276 (1956) — S.B. Roy

Vibrational Analysis — Spectrochim. Acta 12, 305 (1958) — C. Garrigou-Lagrange, J.M. Lebas and M.L. Josien

Vibrational Analysis — J. Chem. Soc. 2051 (1959) — A.R. Katritzky and P. Simons

Fluorescence — Indian J. Phys. 33, 209 (1959) — J.K. Roy

UV Absorption in Liquid — Indian J. Phys. 34, 331 (1960) — J.K. Roy

UV Absorption in Vapour — Indian J. Phys. 35, 143 (1961) — J.K. Roy

UV Absorption in Solution — Indian J. Phys. 36, 156 (1962) — J.K. Roy

Vibrational Analysis — Spectrochim. Acta 20, 1343 (1964) — E.F. Mooney

IR Spectrum — Indian J. Phys. 39, 537 (1965) — D.K. Mukharjee, P.K. Bishnui and S.C. Sirkar

Raman — Indian J. Phys. <u>40</u>, 415 (1966) —
   S.K. Nandy
Vibrational Analysis — J. Sci. Res. BHU
   (India) <u>20(2)</u>, 25 (1970) — R. Amni Amma,
   K.P.R. Nair and D.K. Rai
Vibrational Analysis — Spectrochim. Acta
   <u>26A</u>, 1503 (1970) — J.H.S. Green
Raman — J. Mol. Struct. <u>5</u>, 477 (1970) — N.T.
   McDevitt and W.G. Fateley
Calculation — J. Mol. Struct. <u>8</u>, 319 (1971)—
   P.C. Mishra, S.N. Thakur and D.K. Rai
Vibrational Analysis — Spectrochim. Acta
   <u>28A</u>, 55 (1972) — L. Verdonck and G.P.
   Van der Kelen

2-Chloro-m-Xylene ($C_6H_3(CH_3)_2Cl$)

Raman — Monatsh. Chem. <u>64</u>, 361 (1934) —
   K.W.F. Kohlrausch and A. Pongratz
Vibrational Analysis — Spectrochim. Acta
   <u>27A</u>, 793 (1971) — J.H.S. Green, D.J.
   Harrison and W. Kynaston

5-Chloro-m-Xylene ($C_6H_3(CH_3)_2Cl$)

Raman — Monatsh. Chem. <u>76</u>, 215 (1946) —
   K.W.F. Kohlrausch
Vibrational Analysis — Spectrochim. Acta
   <u>27A</u>, 793 (1971) — J.H.S. Green, D.J.
   Harrison and W. Kynaston

4-Chloro-o-Xylene ($C_6H_3(CH_3)_2Cl$)

IR Spectrum — Indian J. Pure Appl. Phys.
   <u>8</u>, 483 (1970) — M.A. Shashidhar, K.S.
   Rao and E.S. Jayadevappa
UV Absorption in Vapour — Indian J. Pure
   Appl. Phys. <u>16</u>, 634 (1978) — P.V.
   Shanbhag, M.A. Shashidhar and K.S. Rao

2-Chloro-p-Xylene ($C_6H_3(CH_3)_2Cl$)

UV Absorption in Vapour — Curr. Sci. <u>35</u>,
   616 (1966) — M.A. Shashidhar and K.S.
   Rao
IR Spectrum — Indian J. Pure Appl. Phys.
   <u>8</u>, 483 (1966) — M.A. Shashidhar, K.S.
   Rao and E.S. Jayadevappa

1-Chloro-2,6-Xylene ($C_6H_3(CH_3)_2Cl$)

IR Spectrum — Indian J. Pure Appl. Phys.
   <u>8</u>, 483 (1970) — M.A. Shashidhar, K.S.
   Rao and E.S. Jayadevappa

UV Absorption in Vapour — Indian J. Pure
   Appl. Phys. <u>16</u>, 634 (1978) — P.V.
   Shanbhag, M.A. Shashidhar and K.S. Rao

Cinnoline ($C_8H_6N_2$)

UV Absorption in Vapour — J. Mol. Spectrosc.
   <u>24</u>, 383 (1967) — S.C. Wait, Jr and
   F.M. Grogan
IR Spectrum — Spectrochim. Acta <u>22</u>, 117
   (1966) — W.L.F. Armarego, G.B. Barlin
   and E. Spinner
Fluorescence — Chem. Phys. Lett. <u>21</u>, 326
   (1973) — J.A. Stikeleather
Absorption Spectrum in Solid — J. Mol.
   Spectrosc. <u>46</u>, 316 (1973) — A.J. Jordan
   and I.G. Ross
Fluorescence — J. Chem. Phys. <u>61</u>, 3895
   (1974) — J.R. McDonald and L.E. Brus
Phosphorescence — J. Chem. Phys. <u>61</u>, 3895
   (1974) — J.R. McDonald and L.E. Brus
Calculation — J. Mol. Spectrosc. <u>50</u>, 457
   (1974) — J.E. Ridley and M.C. Zerner

Cinnoline-D_6 ($C_8D_6N_2$)

Absorption Spectrum — J. Mol. Spectrosc.
   <u>49</u>, 201 (1974) — G. Fischer

Collidine ($C_5H_2N(CH_3)_3$)

See   2,4,6-Trimethylpyridine

Coumarin ($C_9H_6O_2$)

Raman — Curr. Sci. <u>6</u>, 328 (1938) — C.S.
   Venkateshwaran
Raman — Curr. Sci. <u>37</u>, 10 (1968) — C.P.
   Girijavallabhan and K. Venkteshwarlu
Absorption Spectrum — J. Phys. Chem. <u>71</u>,
   4234 (1970) — P.S. Song and W.H. Gordon
Absorption Spectrum — Photochem. Photobiol.
   <u>14</u>, 521 (1971) — P.S. Song, M.L.
   Haster, T.A. Moore and W.C. Herndon
Absorption Spectrum — J. Mol. Spectrosc.
   <u>39</u>, 466 (1971) — B.R. Henry and R.V.
   Hunt
Absorption Spectrum — J. Mol. Spectrosc.
   <u>40</u>, 144 (1971) — T.A. Moore, M.L.
   Haster and P.S. Song
Fluorescence — J. Mol. Spectrosc. <u>40</u>, 144
   (1971) — T.A. Moore, M.L. Haster and
   P.S. Song
Absorption Spectrum — J. Mol. Spectrosc.
   <u>48</u>, 117 (1973) — B.R. Henry and E.A.
   Lawler

o-Cresol ($C_6H_4OHCH_3$)

UV Absorption in Vapour – Ann. Chim.(Paris)
21, 287 (1929) – J. Savard
Raman – Monatsh. Chem. 76, 1 (1946) – E.
Herz
UV Absorption in Different States – Indian
J. Phys. 26, 119 (1952) – H.N. Swamy
Raman – Indian J. Phys. 29, 257 (1955) –
D.C. Biswas
Fluorescence – Indian J. Phys. 29, 257
(1955) – D.C. Biswas
Vibrational Analysis – Spectrochim. Acta
18, 1433 (1962) – K. Venketeshwar and
M. Radhakrishnan
Vibrational Analysis – Chem. Ind.(London) –
1575 (1962) – J.H.S. Green, W. Kynaston
and H.A. Gebbie
UV Absorption in Vapour – Proc. Indian Acad.
Sci. 55, 247 (1962) – P.R. Rao
Vibrational Analysis – J. Chim. Phys. 62,
347 (1965) – M. Brigodiot and J.M.
Lebas
Calculation – Opt. Spektrosk. 18, 605
(1965) – N.I. Davydova, I.A. Zhigunova,
L.A. Ignatyeva and M.A. Kovner
Vibrational Analysis – Vestnik. Moskov Univ.
4, 39 (1968) – N.I. Davydova, I.A.
Zhigunova, L.A. Ignatyeva, M.A. Kovner
and A.Y. Siderdu
IR Spectrum – Indian J. Phys. 44, 9 (1970) –
J. Jha and S. Chattopadhyay
Vibrational Analysis – Spectrochim. Acta
27A, 2199 (1971) – J.H.S. Green, D.J.
Harrison and W. Kynaston

o-Cresol-OD ($C_6H_4ODCH_3$)

Vibrational Analysis – Spectrochim. Acta
27A, 2199 (1971) – J.H.S. Green, D.J.
Harrison and W. Kynaston

m-Cresol ($C_6H_4OHCH_3$)

UV Absorption in Vapour – Ann. Chim.(Paris)–
21, 287 (1929) – J. Savard
Raman – Monatsh. Chem. 74, 160 (1942) –
E. Herz
UV Absorption in Different States – Indian
J. Phys. 26, 119 (1952) – H.N. Swamy
Raman – Indian J. Phys. 29, 257 (1955) –
D.C. Biswas
Fluorescence – Indian J. Phys. 29, 257
(1955) – D.C. Biswas
Vibrational Analysis – Spectrochim. Acta
18, 1433 (1962) – K. Venketeshwar and
M. Radhakrishnan

UV Absorption in Vapour – Proc. Indian
Acad. Sci. 55, 300 (1962) – P.R. Rao
Vibrational Analysis – Chem. Ind. (London)
1575 (1962) – J.H.S. Green, W. Kynaston
and H.A. Gebbie
Vibrational Analysis – Spectrochim. Acta
21, 1753 (1965) – R.J. Jackobsen
Vibrational Analysis – J. Chim. Phys. 63,
552 (1966) – C. Garrigou-Lagrange,
M. Chehata and J. Lascombe
Vibrational Analysis – Vestnik. Moskov
Univ. 4, 39 (1968) – N.I. Davydova,
I.A. Zhigunova, L.A. Ignatyeva, M.A.
Kovner and A.Y. Siderdu
IR Spectrum – Indian J. Phys. 44, 9
(1970) – J. Jha and S. Chattopadhyay

p-Cresol ($C_6H_4OHCH_3$)

UV Absorption in Vapour – Ann. Chim.(Paris)
21, 287 (1929) – J. Savard
Raman – Monatsh. Chem. 63, 427 (1934) –
K.W.F. Kohlrausch and A. Pongratz
UV Absorption in Vapour – J. Chem. Phys.
10, 672 (1942) – H. Sponer
UV Absorption in Vapour – Curr. Sci. 20,
201 (1951) – K. Sreeramamurty
UV Absorption in Vapour – Trans. Faraday
Soc. 47A, 1256 (1951) – K. Sreerama-
murty
UV Absorption in Solution – Bull. Chem. Soc.
Jap. 25, 153 (1953) – S. Imanishi and
M. Ito
UV Absorption in Different States – Indian
J. Phys. 26, 119 (1952) – H.N. Swamy
UV Absorption in Vapour – J. Chem. Phys.
20, 532 (1955) – S. Imanishi, M. Ito,
K. Semba and T. Anno
Raman – Indian J. Phys. 29, 257 (1955) –
D.C. Biswas
Fluorescence – Indian J. Phys. 29, 257
(1955) – D.C. Biswas
Vibrational Analysis – Spectrochim. Acta
7, 253 (1955) – D.H. Whiffen
Vibrational Analysis – Spectrochim. Acta
12, 305 (1958) – C. Garrigou-Lagrange,
J.M. Lebas and M.L. Josien
Vibrational Analysis – Chem. Ind. (London)–
1575 (1962) – J.H.S. Green, W. Kynaston
and H.A. Gebbie
Vibrational Analysis – Appl. Spectrosc. 16,
32 (1962) – R.J. Jackobsen and E.J.
Brewer
UV Absorption in Vapour – Proc. Indian
Acad. Sci. 55, 232 (1962) – P.R. Rao
Calculation – Opt. Spektrosk. 18, 605
(1965) – N.I. Davydova, I.A. Zhigu-
nova, I.A. Igenatyeva and M.A. Kovner

Vibrational Analysis — Spectrochim. Acta
21, 1753 (1965) — R.J. Jackobsen and
J. W. Brasch
Vibrational Analysis — Spectrochim. Acta
21, 433 (1965) — R.J. Jackobsen
Vibrational Analysis — J. Amer. Chem. Soc.
88, 3199 (1966) — W.J. Hurley, I.D.
Kuntz and G.E. Leroi
Vibrational Analysis — Appl. Spectrosc. 22,
641 (1968) — R.J. Jackobsen, J.W.
Brasch and Y. Mikawa
Vibrational Analysis — Vestnik. Moskov
Univ. 4, 39 (1968) — N.I. Davydova,
I.A. Zhigugova, L.A. Ignatyeva, M.A.
Kovner and A.Y. Sideridu
IR Spectrum — Indian J. Phys. 44, 9 (1970)—
J. Jha and S. Chattopadhyay
Vibrational Analysis — Spectrochim. Acta
27A, 2199 (1971) — J.H.S. Green, D.J.
Harrison and W. Kynaston

p-Cresol-OD ($C_6H_4ODCH_3$)

Vibrational Analysis — Spectrochim. Acta
21, 433 (1965) — R.J. Jackobsen

p-Cresol-2,6-D_2 ($C_6H_2D_2OHCH_3$)

Vibrational Analysis — Spectrochim. Acta
21, 433 (1965) — R.J. Jackobsen

p-Cresol-3,5-D_2 ($C_6H_2D_2OHCH_3$)

Vibrational Analysis — Spectrochim. Acta
21, 433 (1965) — R.J. Jackobsen

p-Cresol-2,3,5,6-D_4 ($C_6D_4OHCH_3$)

Vibrational Analysis — Spectrochim. Acta
21, 433 (1965) — R.J. Jackobsen
Vibrational Analysis — Vestnik. Moskov
Univ. 4, 39 (1968) — N.I. Davydova,
I.A. Zhigunova, L.A. Ignatyeva, M.A.
Kovner and A.Y. Sideridu

o-Cyanopyridine (C_5H_4NCN)

UV Absorption in Solution — Boll. Sci. Fac.
Chim. Ind. Bologna 15, 90 (1957) — G.
Leandri and D. Spinneli
UV Absorption in Solution — J. Chem. Soc.
3855 (1963) — E. Spinner
Phosphorescence — J. Amer. Chem. Soc. 91,
6508 (1969) — R.J. Hoover and M. Kasha

UV Absorption in Vapour — Curr. Sci. 40,
628 (1971) — M.R. Padhye and A.K.
Gowardhan
UV Absorption in Different States — J.
Mol. Spectrosc. 60, 412 (1976) — D.
Marjit and S.B. Banerjee
Vibrational Analysis — Spectrochim. Acta
33A, 75 (1977) — J.H.S. Green and
D.J. Harrison

m-Cyanopyridine (C_5H_4NCN)

UV Absorption in Solution — Boll. Sci. Fac.
Chim. Ind. Bologna 15, 90 (1957) —
G. Leandri and D. Spinneli
UV Absorption in Solution — J. Chem. Soc.
3855 (1963) — E. Spinner
UV Absorption in Vapour — Curr. Sci. 40,
628 (1971) — M.R. Padhye and A.K.
Gowardhan
UV Absorption in Different States — J. Mol.
Spectrosc. 60, 412 (1976) — D. Marjit
and S.B. Banerjee
Vibrational Analysis — Spectrochim. Acta
33A, 75 (1977) — J.H.S. Green and
D.J. Harrison

p-Cyanopyridine (C_5H_4NCN)

UV Absorption in Solution — Boll. Sci.
Fac. Chim. Ind. Bologna 15, 90
(1957) — G. Leandri and D. Spinneli
UV Absorption in Solution — J. Chem. Soc.
3855 (1963) — E. Sinner
Phosphorescence — J. Amer. Chem. Soc. 91,
6508 (1969) — R.J. Hoover and M. Kasha
UV Absorption in Vapour — Curr. Sci. 40,
628 (1971) — M.R. Padhye and A.K.
Gowardhan
UV Absorption in Different States — J. Mol.
Spectrosc. 60, 412 (1976) — D. Marjit
and S.B. Banerjee
Vibrational Analysis — Spectrochim. Acta
33A, 75 (1977) — J.H.S. Green and
D.J. Harrison

Cyanotoluene ($C_6H_4CNCH_3$)

See   Tolunitrile

Cyclohexylbenzene ($C_6H_5C_6H_5$)

IR Spectrum — Indian J. Phys. 36, 543
(1962) — R.N. Bapat

UV Absorption in Vapour — Indian J. Phys.
    $\underline{41}$, 587 (1967) — R.N. Bapat
Raman — Indian J. Phys. $\underline{44}$, 273 (1970) —
    R.N. Bapat

Diaminobenzene $(C_6H_4(NH_2)_2)$

See   Phenylenediamine

2,3-Diaminopyridine $(C_5H_3N(NH_2)_2)$

UV Absorption in Solution — Spectrochim.
    Acta $\underline{23A}$, 89 (1967) — G. Favini, A.
    Gamba and I.R. Bellobono

2,5-Diaminopyridine $(C_5H_3N(NH_2)_2)$

UV Absorption in Solution — Spectrochim.
    Acta $\underline{23A}$, 89 (1967) — G. Favini, A.
    Gamba and I.R. Bellobono

2,6-Diaminopyridine $(C_5H_3N(NH_2)_2)$

UV Absorption in Solution — Spectrochim.
    Acta $\underline{23A}$, 89 (1967) — G. Favini, A.
    Gamba and I.R. Bellobono
UV Absorption in Vapour — J. Chim. Phys.
    $\underline{65}$, 1486 (1968) — M. Lamotte and P.
    Loustauneau
Vibrational Analysis — Indian J. Pure Appl.
    Phys. $\underline{8}$, 761 (1970) — G.D. Baruah, R.
    Amni Amma, P.S. Dube and S.N. Rai

3,4-Diaminopyridine $(C_5H_3N(NH_2)_2)$

UV Absorption in Solution — Spectrochim.
    Acta $\underline{23A}$, 89 (1967) — G. Favini, A.
    Gamba and I.R. Bellobono

1,2-Diazanaphthalene $(C_8H_6N_2)$

See   Cinnoline

1,3-Diazanaphthalene $(C_8H_6N_2)$

See   Quinazoline

1,4-Diazanaphthalene $(C_8H_6N_2)$

See   Quinoxaline

1,5-Diazanaphthalene $(C_8H_6N_2)$

IR Spectrum — Spectrochim. Acta $\underline{22}$, 117
    (1966) — W.L.F. Armarego, G.B. Barlin
    and E. Spinner

1,6-Diazanaphthalene $(C_8H_6N_2)$

IR Spectrum — Spectrochim. Acta $\underline{22}$, 117
    (1966) — W.L.F. Armarego, G.B. Barlin
    and E. Spinner

1,8-Diazanaphthalene $(C_8H_6N_2)$

IR Spectrum — Spectrochim. Acta $\underline{22}$, 117
    (1966) — W.L.F. Armarego, G.B. Barlin
    and E. Spinner

2,3-Diazanaphthalene $(C_8H_6N_2)$

See   Phthalazine

1,2-Diazine $(C_4H_4N_2)$

See   Pyridazine

1,4-Diazine $(C_4H_4N_2)$

See   Pyrazine

2,4-Dibromoaniline $(C_6H_3NH_2Br_2)$

IR Spectrum — J. Indian Chem. Soc. $\underline{47}$,
    1169 (1970) — V.N. Verma, K.P.R. Nair
    and D.K. Rai
UV Absorption in Vapour — Appl. Spectrosc.
    $\underline{26}$, 107 (1972) — V.N. Verma, K.P.R.
    Nair and D.K. Rai
UV Absorption in Solution — J. Sci. Res.
    BHU (India) $\underline{23(2)}$, 175 (1972) —
    V.N. Verma

2,5-Dibromoaniline $(C_6H_3NH_2Br_2)$

IR Spectrum — J. Indian Chem. Soc. $\underline{47}$,
    1169 (1970) — V.N. Verma, K.P.R. Nair
    and D.K. Rai
UV Absorption in Solution — J. Sci. Res.
    BHU (India) $\underline{23(2)}$, 175 (1972) —
    V.N. Verma
UV Absorption in Vapour — Indian J. Pure
    Appl. Phys. $\underline{13}$, 421 (1975) — V.N.
    Verma

2,6-Dibromoaniline ($C_6H_3NH_2Br_2$)

IR Spectrum — J. Indian Chem. Soc. <u>47</u>, 1169 (1970) — V.N. Verma, K.P.R. Nair and D.K. Rai
Vibrational Analysis — Spectrosc. Lett. <u>5</u>, 133 (1972) — V.N. Verma
UV Absorption in Solution — J. Sci. Res. BHU (India) <u>23(2)</u>, 175 (1972) — V.N. Verma

o-Dibromobenzene ($C_6H_4Br_2$)

Raman — Wien. Ber. <u>144</u>, 417 (1935) — K.W.F. Kohlrausch and G.P. Ypsilanti
Raman — Monatsh.Chem. <u>66</u>, 285 (1935) — K.W.F. Kohlrausch and Gr. Prinz
Raman — Monatsh. Chem. <u>76</u>, 1 (1946) — E. Herz
Vibrational Analysis — Spectrochim. Acta <u>19</u>, 807 (1961) — J.H.S. Green, W. Kynaston and H.A. Gebbie
Vibrational Analysis — J. Chim. Phys. <u>62</u>, 347 (1965) — M. Brigodiot and J.M. Lebas
IR Spectrum — Spectrochim. Acta <u>22</u>, 333 (1966) — H.F. Shurvell, B. Dulaurens and P. Pesteil
Vibrational Analysis — Spectrochim. Acta <u>24A</u>, 853 (1968) — J.H.S. Green, W. Kynaston and G.A. Rodley
UV Absorption in Vapour — Appl. Spectrosc. <u>22</u>, 588 (1968) — Y.P. Shrivastava
Vibrational Analysis — Indian J. Phys. <u>42</u>, 1 (1968) — S.C. Sirkar and P.K. Bishnui
Vibrational Analysis — Spectrochim. Acta <u>26A</u>, 1913 (1970) — J.H.S. Green

m-Dibromobenzene ($C_6H_4Br_2$)

Raman — Wien. Ber. <u>144</u>, 417 (1935) — K.W.F. Kohlrausch and G.P. Ypsilanti
Raman — Monatsh. Chem. <u>66</u>, 285 (1935) — K.W.F. Kohlrausch and Gr. Prinz
Vibrational Analysis — Spectrochim. Acta <u>22</u>, 333 (1966) — H.F. Shurvell, B. Dulaurens and P. Pesteil
Vibrational Analysis — Indian J. Phys. <u>42</u>, 1 (1968) — S.C. Sirkar and P.K. Bishnui
Vibrational Analysis — Spectrochim. Acta <u>26A</u>, 1523 (1970) — J.H.S. Green

p-Dibromobenzene ($C_6H_4Br_2$)

Raman — Monatsh. Chem. <u>66</u>, 285 (1935) — K.W.F. Kohlrausch and Gr. Prinz

Raman — Wien. Ber. <u>144</u>, 417 (1935) — K.W.F. Kohlrausch and G.P. Ypsilanti
UV Absorption in Vapour — Curr. Sci. <u>20</u>, 176, 291 (1951) — K. Sreeramamurty
Vibrational Analysis — Dokl. Akad. Nauk. SSSR <u>41</u>, 499 (1953) — M.A. Kovner
Vibrational Analysis — Spectrochim. Acta <u>12</u>, 47 (1958) — A. Stojilkovic and D.H. Whiffen
UV Absorption (Crystal) — Indian J. Phys. <u>33</u>, 276 (1959) — T.N. Misra
Vibrational Analysis — Spectrochim. Acta <u>22</u>, 333 (1966) — H.F. Shurvell, B. Dulaurens and P. Pesteil
Vibrational Analysis — J. Chim. Phys. <u>64</u>, 1450 (1967) — P. Pommez, M. Lafaix, P. Delorme and V. Lorenzelli
Raman — Spectrochim. Acta <u>25A</u>, 1017 (1969) — M. Sukuki and M. Ito
Vibrational Analysis — Spectrochim. Acta <u>26A</u>, 1503 (1970) — J.H.S. Green

p-Dibromobenzene-D_4 ($C_6D_4Br_2$)

Vibrational Analysis — Spectrochim. Acta <u>25A</u>, 507 (1969) — P.N. Gates, K. Radcliffe and D. Steele
UV Absorption in Vapour — Appl. Spectrosc. <u>25</u>, 85 (1971) — J.V. Shukla, K.N. Upadhya and S.N. Thakur

2,4-Dibromofluorobenzene ($C_6H_3FBr_2$)

Vibrational Analysis — Indian J. Pure Appl. Phys. <u>15</u>, 432 (1977) — G.N.R. Tripathi and V.M. Pandey

2,4-Dibromophenol ($C_6H_3OHBr_2$)

IR Spectrum — Indian J. Pure Appl. Phys. <u>12</u>, 597 (1974) — B.K. Dwivedi, I.D. Singh and R.C. Maheshwari

2,6-Dibromophenol ($C_6H_3OHBr_2$)

IR Spectrum — Indian J. Pure Appl. Phys. <u>12</u>, 597 (1974) — B.K. Dwivedi, I.D. Singh and R.C. Maheshwari

2,5-Dibromopyridine ($C_5H_3NBr_2$)

UV Absorption in Vapour — Rec. Trav. Chim. Pays-Bas <u>56</u>, 573 (1937) — C.W.F. Spiers and J.P. Wibaut

UV Absorption in Different States — Indian
  J. Phys. <u>47</u>, 480 (1973) — D. Marjit
Vibrational Analysis — Spectrochim. Acta
  <u>29A</u>, 1177 (1973) — J.H.S. Green, D.J.
  Harrison and M.R. Kipps

2,6-Dibromopyridine $(C_5H_3NBr_2)$

UV Absorption in Different States — Indian
  J. Phys. <u>47</u>, 480 (1973) — D. Marjit
Vibrational Analysis — Spectrochim. Acta
  <u>29A</u>, 1177 (1973) — J.H.S. Green, D.J.
  Harrison and M.R. Kipps

3,5-Dibromopyridine $(C_5H_3NBr_2)$

Vibrational Analysis — Spectrochim. Acta
  <u>29A</u>, 1177 (1973) — J.H.S. Green, D.J.
  Harrison and M.R. Kipps

p-Dibromotetradeuterobenzene $(C_6D_4Br_2)$

Vibrational Analysis — Spectrochim. Acta
  <u>25A</u>, 507 (1969) — P.N. Gates, K.
  Radcliffe and D. Steele

1,2-Dibromotetrafluorobenzene $(C_6F_4Br_2)$

Vibrational Analysis — Spectrochim. Acta
  <u>33A</u>, 193 (1977) — J.H.S. Green and
  D.J. Harrison

1,3-Dibromotetrafluorobenzene $(C_6F_4Br_2)$

Vibrational Analysis — Spectrochim. Acta
  <u>33A</u>, 193 (1977) — J.H.S. Green and
  D.J. Harrison

1,4-Dibromotetrafluorobenzene $(C_6F_4Br_2)$

Vibrational Analysis — Spectrochim. Acta
  <u>33A</u>, 193 (1977) — J.H.S. Green and
  D.J. Harrison

2,5-Dibromotoluene $(C_6H_3CH_3Br_2)$

Vibrational Analysis — Bull. Soc. Chim. Fr.
  <u>13</u>, 540 (1946) — R. Pajeau

3,4-Dibromotoluene $(C_6H_3CH_3Br_2)$

Vibrational Analysis — Bull. Soc. Chim. Fr.
  <u>13</u>, 540 (1946) — R. Pajeau

3,5-Dibromotrifluoropyridine $(C_5F_3Br_2N)$

Calculation — Spectrochim. Acta <u>19</u>, 1791
  (1963) — D.A. Long and D. Steele
Vibrational Analysis — Spectrochim. Acta
  <u>33A</u>, 81 (1977) — J.H.S. Green and
  D.J. Harrison

2,4-Dichloroaniline $(C_6H_3NH_2Cl_2)$

UV Absorption in Vapour — Indian J. Pure
  Appl. Phys. <u>5</u>, 305 (1967) — D. Sharma
  and S.L. Shrivastava
IR Spectrum — Indian J. Pure Appl. Phys.
  <u>5</u>, 189 (1967) — S.L. Shrivastava
UV Absorption in Vapour — Indian J. Phys.
  <u>52B</u>, 125 (1978) — M.A. Shashidhar

2,5-Dibromoaniline $(C_6H_3NH_2Cl_2)$

Raman — Monatsh. Chem. <u>67</u>, 80 (1936) —
  K.W.F. Kohlrausch, W. Stockmair and
  G.P. Ypsilanti
Raman — Monatsh. Chem. <u>75</u>, 49 (1947) —
  K.W.F. Kohlrausch, E. Herz and R.
  Vogel
Vibrational Analysis — Proc. Roy. Soc.
  <u>A243</u>, 143 (1957) — P.J. Krueger and
  H.W. Thompson
Vibrational Analysis — Can. J. Chem. <u>40</u>,
  2300 (1962) — P.J. Krueger
IR Spectrum — Aust. J. Chem. <u>15</u>, 626
  (1962) — A.N. Hamby and B.V.O. Grady
UV Absorption in Vapour — Curr. Sci. <u>35</u>,
  62 (1966) — S.N. Singh and N.L. Singh
UV Absorption in Vapour — Indian J. Pure
  Appl. Phys. <u>5</u>, 305 (1967) — D. Sharma
  and S.L. Shrivastava
IR Spectrum — Indian J. Pure Appl. Phys.
  <u>5</u>, 189 (1967) — S.L. Shrivastava
IR Spectrum — Indian J. Pure Appl. Phys.
  <u>7</u>, 250 (1969) — S.N. Singh and N.L.
  Singh

2,6-Dichloroaniline $(C_6H_3NH_2Cl_2)$

IR Spectrum — Indian J. Pure Appl. Phys.
  <u>7</u>, 250 (1969) — S.N. Singh and N.L.
  Singh

UV Absorption in Vapour — Indian J. Pure
   Appl. Phys. $\underline{8}$, 764 (1970) — C.G. Rama
   Rao, B.R.K. Reddy and P.T. Rao
UV Absorption in Vapour — Indian J. Pure
   Appl. Phys. $\underline{9}$, 138 (1971) — S. Nath
   Singh and S.N. Singh
Raman — Indian J. Pure Appl. Phys. $\underline{10}$, 637
   (1972) — P.K. Bishnui

3,4-Dichloroaniline $(C_6H_3NH_2Cl_2)$

UV Absorption in Vapour — Indian J. Pure
   Appl. Phys. $\underline{8}$, 764 (1970) — C.G. Rama
   Rao, B.R.K. Reddy and P.T. Rao

1,4-Dichloroanthraquinone$(C_{14}H_6Cl_2O_2)$

Emission Spectrum in Vapour — Indian J.
   Pure Appl. Phys. $\underline{7}$, 814 (1969) — S. Nath
   Singh, G.D. Baruah and R.S. Singh

1,5-Dichloroanthraquinone$(C_{14}H_6Cl_2O_2)$

IR Spectrum — Indian J. Pure Appl. Phys.
   $\underline{6}$, 454 (1968) — S. Nath Singh and R.S.
   Singh
Emission Spectrum in Vapour — Indian J. Pure
   Appl. Phys. $\underline{7}$, 814 (1969) — S. Nath
   Singh, G.D. Baruah and R.S. Singh
UV Absorption in Vapour — Indian J. Pure
   Appl. Phys. $\underline{8}$, 641 (1970) — S. Nath
   Singh and R.S. Singh

1,8-Dichloroanthraquinone$(C_{14}H_6Cl_2O_2)$

IR Spectrum — Indian J. Pure Appl. Phys. $\underline{6}$,
   454 (1968) — S. Nath Singh and R.S.
   Singh
Emission Spectrum in Vapour — Indian J.
   Pure Appl. Phys. $\underline{7}$, 814 (1969) — S. Nath
   Singh, G.D. Baruah and R.S. Singh
UV Absorption in Vapour — Indian J. Pure
   Appl. Phys. $\underline{8}$, 641 (1970) — S. Nath
   Singh and R.S. Singh

2,4-Dichlorobenzaldehyde $(C_6H_3CHOCl_2)$

UV Absorption in Solution — J. Amer. Chem.
   Soc. $\underline{77}$, 4335 (1955) — L. Doub and J.M.
   Vanderbelt
IR Spectrum — Indian J. Pure Appl. Phys.
   $\underline{10}$, 545 (1972) — H.S. Singh and N.K.
   Sanyal

2,6-Dichlorobenzaldehyde $(C_6H_3CHOCl_2)$

IR Spectrum — Anal. Chem. $\underline{27}$, 2 (1955) —
   S. Pinchas
IR Spectrum — Anal. Chem. $\underline{29}$, 334 (1957) —
   S. Pinchas
IR Spectrum — Indian J. Pure Appl. Phys.
   $\underline{10}$, 545 (1972) — H.S. Singh and N.K.
   Sanyal

3,4-Dichlorobenzaldehyde $(C_6H_3CHOCl_2)$

IR Spectrum — Indian J. Pure Appl. Phys.
   $\underline{10}$, 545 (1972) — H.S. Singh and N.K.
   Sanyal

o-Dichlorobenzene $(C_6H_4Cl_2)$

Raman — Monatsh. Chem. $\underline{61}$, 426 (1932) —
   K.W.F. Kohlrausch, A. Dadiev and A.
   Pongratz
Raman — J. Chem. Phys. $\underline{9}$, 667 (1941) — H.
   Sponer and J.S. Smith-Kirby
UV Absorption (Crystal) — J. Chem. Phys.
   $\underline{20}$, 1177 (1952) — S.C. Sirkar and H.N.
   Swamy
UV Absorption in Vapour — J. Chem. Soc.
   $\underline{23}$, 796 (1953) — T. Anno and I. Matub-
   ara
UV Absorption in Solid — Indian J. Phys.
   $\underline{27}$, 55 (1953) — H.N. Swamy
UV Absorption in Vapour — J. Chem. Phys.
   $\underline{22}$, 234 (1954) — H. Sponer
Raman — Indian J. Phys. $\underline{29}$, 179 (1955) —
   D.C. Biswas
Raman — Indian J. Phys. $\underline{29}$, 558 (1955) —
   D.C. Biswas
UV Absorption in Vapour — Indian J. Phys.
   $\underline{31}$, 444 (1957) — S.L. N.G. Krishnama-
   chari
UV Absorption in Solution — Indian J. Phys.
   $\underline{31}$, 177 (1957) — S.B. Roy and S.C.
   Sirkar
Vibrational Analysis — Spectrochim. Acta
   $\underline{18}$, 1433 (1962) — K. Venkateshwarlu
   and M. Radhakrishnan
Vibrational Analysis — Spectrochim. Acta
   $\underline{19}$, 173 (1963) — J.R. Scherer and J.C.
   Evans
IR Spectrum — J. Sci. Res. BHU (India) —
   $\underline{15}$, 283 (1964) — I.S. Singh, V.B. Singh
   and V.N. Sharma
UV Absorption in Different States — Indian
   J. Pure Appl. Phys. $\underline{7}$, 62 (1969) —
   K.G. Srinivasacharya
UV Absorption in Vapour — Mol. Phys. $\underline{18}$,
   261 (1970) — T. Cvitas and J.M. Hollas

Raman — J. Mol. Struct. 5, 477 (1970) —
   N.T. McDevitt and W.G. Fateley
Vibrational Analysis — Spectrochim. Acta
   26A, 1913 (1970) — J.H.S. Green

o-Dichlorobenzene-D₄ (C₆D₄Cl₂)

Vibrational Analysis — Spectrochim. Acta
   19, 1739 (1963) — J.R. Scherer and
   J.C. Evans

m-Dichlorobenzene (C₆H₄Cl₂)

Raman — J. Chem. Phys. 1, 512 (1933) — J.W.
   Swaine and J.W. Murray
IR Spectrum — J. Phys. Radium 9, 13 (1938) —
   J. Lecomte
Raman — J. Chem. Phys. 9, 667 (1941) — H.
   Sponer and J.S. Smith-Kirby
UV Absorption in Vapour — J. Chem. Phys.
   22, 234 (1954) — H. Sponer
Raman — Indian J. Phys. 29, 179 (1955) —
   D.C. Biswas
UV Absorption in Solid — Indian J. Phys.
   30, 242 (1956) — G.S.R. Krishnamurty
   and S.N. Sen
UV Absorption in Solution — Indian J. Phys.
   31, 177 (1957) — S.B. Roy and S.C.
   Sirkar
UV Absorption in Different States — Indian
   J. Phys. 31, 99 (1957) — S.K. Sen
UV Absorption in Vapour — Indian J. Phys.
   31, 447 (1957) — S.L.N.G. Krishnama-
   chari
Vibrational Analysis — Spectrochim. Acta
   19, 1754 (1963) — J.R. Scherer and
   J.C. Evans
Vibrational Analysis — J. Chim. Phys. 63,
   552 (1966) — C. Garrigou-Lagrange, M.
   Chehata and J. Lascombe
UV Absorption in Different States — Indian
   J. Pure Appl. Phys. 7, 62 (1969) —
   K.G. Srinivasacharya
Vibrational Analysis — Spectrochim. Acta
   26A, 1523 (1970) — J.H.S. Green
Raman — J. Mol. Struct. 5, 477 (1970) —
   N.T. McDevitt and W.G. Fateley
UV Absorption in Vapour — Indian J. Phys.
   51B, 455 (1977) — S.J. Singh and S.M.
   Pandey

p-Dichlorobenzene (C₆H₄Cl₂)

UV Absorption in Solution — Z. Phys. Chem.
   B19, 76 (1932) — H. Conrad-Billroth

Raman — J. Chem. Phys. 1, 512 (1933) —
   J.W. Swaine and J.W. Murray
Raman — Indian J. Phys. 10, 473 (1936) —
   S.C. Sirkar and P. Gupta
Raman — Indian J. Phys. 11, 287 (1937) —
   S.C. Sirkar and P. Gupta
Raman — Indian J. Phys. 11, 418 (1937) —
   S.C. Sirkar and P.K. Bishnui
Raman — Indian Acad. Sci. 8, 448 (1938) —
   G. Venkateshwaren
Raman — Monatsh. Chem. 72, 268 (1939) —
   O. Paulsen and K.W.F. Kohlrausch
UV Absorption in Vapour — Rev. Mod. Phys.
   13, 76 (1941) — H. Sponer
Raman — J. Chem. Phys. 9, 667 (1941) — H.
   Sponer and J.S. Smith-Kirby
UV Absorption in Vapour — Rev. Mod. Phys.
   14, 224 (1942) — H. Sponer
Raman — Monatsh. Chem. 76, 200 (1947) —
   K.W.F. Kohlrausch and E. Herz
Raman — Indian J. Phys. 25, 79 (1951) —
   H. Narain and B.D. Saksena
Raman — Indian J. Phys. 25, 459 (1951) —
   A.K. Ray
Vibrational Analysis — Dokl. Akad. Nauk.
   SSSR 41, 499 (1953) — M.A. Kovner
UV Absorption in Vapour — J. Chem. Phys.
   22, 234 (1954) — H. Sponer
UV Absorption in Vapour — J. Chem. Phys.
   23, 796 (1955) — T. Anno and I.
   Matubara
UV Absorption in Vapour — Indian J. Phys.
   31, 447 (1957) — S.L.N.G. Krishnama-
   chari
UV Absorption in Solution — Indian J. Phys.
   31, 177 (1957) — S.B. Roy and S.C.
   Sirkar
Vibrational Analysis — Spectrochim. Acta
   12, 305 (1958) — C. Garrigou-Lagrange,
   J.M. Lebas and M.L. Josien
IR Spectrum — Spectrochim. Acta 12, 47
   (1958) — A. Stojilkovic and D.H.
   Whiffen
UV Absorption (Crystal) — Indian J. Phys.
   33, 45 (1959) — S.C. Sirkar and T.N.
   Misra
IR Spectrum — Spectrochim. Acta 16, 58
   (1960) — E.R. Lippincott, C.E. Weir,
   A. Van Valkenburg and E.N. Bunting
Raman — Bull. Chem. Soc. Jap. 34, 1658
   (1961) — S. Saeki
Calculation — Spectrochim. Acta 19, 345
   (1963) — J.R. Scherer
Vibrational Analysis — Spectrochim. Acta
   19, 1739 (1963) — J.R. Scherer and
   J.C. Evans
Emission Spectrum in Vapour — Indian J.
   Pure Appl. Phys. 2, 284 (1964) —
   K.N. Upadhya and J.N. Rai

Fluorescence — Indian J. Pure Appl. Phys.
   $\underline{2}$, 284 (1964) — K.N. Upadhya and J.N.
   Rai
Vibrational Analysis — Proc. Roy. Soc.
   $\underline{A298}$, 51 (1967) — P.R. Griffiths and
   H.W. Thompson
Calculation — Spectrochim. Acta $\underline{23}$, 1489
   (1967) — J.R. Scherer
UV Absorption in Vapour — J. Chem. Phys.
   $\underline{46}$, 3617 (1967) — G. Castro and R.M.
   Hochstrasser
UV Absorption in Vapour — Proc. Roy. Soc.
   $\underline{A307}$, 97 (1968) — J. Christoffersen,
   J.M. Hollas and G.H. Kirby
UV Absorption in Different States — Indian
   J. Pure Appl. Phys. $\underline{7}$, 62 (1969) —
   K.G. Srinivasacharya
Raman — Spectrochim. Acta $\underline{25A}$, 1017 (1969)—
   M. Sukuki and M. Ito
Vibrational Analysis — Spectrochim. Acta
   $\underline{26A}$, 1503 (1970) — J.H.S. Green
IR Spectrum (Crystal) — Spectrochim. Acta
   $\underline{26A}$, 1603 (1970) — K.M.M. Kruso
UV Absorption in Vapour (High Resolution) —
   Mol. Phys. $\underline{18}$, 801 (1970) — T. Cvitas
   and J.M. Hollas

p-Dichlorobenzene-D_4 ($C_6D_4Cl_2$)

Raman — Bull. Chem. Soc. Jap. $\underline{34}$, 1658
   (1961) — S. Saeki
Vibrational Analysis — Spectrochim. Acta
   $\underline{19}$, 1739 (1963) — J.R. Scherer and
   J.C. Evans
Vibrational Analysis — Spectrochim. Acta
   $\underline{25A}$, 507 (1969) — P.N. Gates, K.
   Radcliffe and D. Steele

2,6-Dichlorobenzonitrile ($C_6H_3CNCl_2$)

UV Absorption in Vapour — Curr. Sci. $\underline{44}$,
   657 (1975) — N.K. Sanyal and P. Ahmed

3,5-Dichlorobenzonitrile ($C_6H_3CNCl_2$)

UV Absorption in Vapour — Curr. Sci. $\underline{44}$,
   657 (1975) — N.K. Sanyal and P. Ahmed

2,3-Dichloro-p-Benzoquinone-D_2 ($C_6D_2Cl_2O_2$)

Vibrational Analysis — Spectrochim. Acta
   $\underline{34A}$, 453 (1978) — A. Girlands and C.
   Pecile

2,5-Dichlorobenzoquinone ($C_6H_2Cl_2O_2$)

UV Absorption in Solution — J. Chem. Soc.
   $\underline{44}$, 490 (1945) — E.A. Braude
Vibrational Analysis — Spectrochim. Acta
   $\underline{20}$, 127 (1964) — F.E. Prichard
UV Absorption in Vapour — Indian J. Pure
   Appl. Phys. $\underline{5}$, 284 (1967) — S.C.
   Srivastava and R.S. Singh
Emission Spectrum in Vapour — Indian J.
   Pure Appl. Phys. $\underline{6}$, 91 (1968) — R.S.
   Singh

2,5-Dichloro-p-Benzoquinone-D_2 ($C_6D_2Cl_2O_2$)

Vibrational Analysis — Spectrochim. Acta
   $\underline{34A}$, 453 (1978) — A. Girlands and
   C. Pecile

2,6-Dichlorobenzoquinone ($C_6H_2Cl_2O_2$)

UV Absorption in Solution — J. Chem. Soc.
   44 & 490 (1945) — E.A. Braude
Vibrational Analysis — Spectrochim. Acta
   $\underline{20}$, 127 (1964) — F.E. Prichard
UV Absorption in Vapour — Indian J. Pure
   Appl. Phys. $\underline{5}$, 284 (1967) — S.C.
   Srivastava and R.S. Singh

2,6-Dichloro-p-Benzoquinone-D_2 ($C_6D_2Cl_2O_2$)

Vibrational Analysis — Spectrochim. Acta
   $\underline{34A}$, 453 (1978) — A. Girlands and
   C. Pecile

2,5-Dichlorobromobenzene ($C_6H_3BrCl_2$)

UV Absorption in Vapour — Indian J. Pure
   Appl. Phys. $\underline{10}$, 489 (1972) — G.N.R.
   Tripathi and V.M. Pandey
Vibrational Analysis — Indian J. Pure
   Appl. Phys. $\underline{15}$, 432 (1977) — G.N.R.
   Tripathi and V.M. Pandey

3,4-Dichlorobromobenzene ($C_6H_3BrCl_2$)

UV Absorption in Vapour — Indian J. Pure
   Appl. Phys. $\underline{10}$, 489 (1972) — G.N.R.
   Tripathi and V.M. Pandey

Vibrational Analysis — Indian J. Pure Appl.
Phys. 15, 432 (1977) — G.N.R. Tripathi
and V.M. Pandey

1,2-Dichloro-4-Fluorobenzene ($C_6H_3Cl_2F$)

Vibrational Analysis — Spectrochim. Acta
27A, 807 (1971) — J.H.S. Green, D.J.
Harrison and W. Kynaston

1,3-Dichloro-2-Fluorobenzene ($C_6H_3Cl_2F$)

Vibrational Analysis — Spectrochim. Acta
27A, 793 (1971) — J.H.S. Green, D.J.
Harrison and W. Kynaston

3,4-Dichlorofluorobenzene ($C_6H_3Cl_2F$)

Vibrational Analysis — Spectrochim. Acta
28A, 33 (1972) — J.H.S. Green, D.J.
Harrison and W. Kynaston
UV Absorption in Vapour — Indian J. Pure
Appl. Phys. 11, 374 (1973) — G.N.R.
Tripathi, V.M. Pandey and S. Ram
Vibrational Analysis — Indian J. Pure Appl.
Phys. 15, 432 (1977) — G.N.R. Tripathi
and V.M. Pandey

2,3-Dichloronaphthoquinone ($C_{10}H_4Cl_2O_2$)

Vibrational Analysis — Indian J. Pure Appl.
Phys. 9, 1071 (1971) — S. Nath Singh,
G.D. Baruah and R.S. Singh

2,3-Dichlorophenol ($C_6H_3OHCl_2$)

IR Spectrum — Indian J. Pure Appl. Phys.
5, 189 (1967) — S.L. Srivastava
Vibrational Analysis — Indian J. Pure Appl.
Phys. 8, 237 (1970) — S.L. Srivastava

2,4-Dichlorophenol ($C_6H_3OHCl_2$)

IR Spectrum — Indian J. Pure Appl. Phys. 5,
189 (1967) — S.L. Srivastava
Vibrational Analysis — Indian J. Pure Appl.
Phys. 8, 237 (1970) — S.L. Srivastava
UV Absorption in Vapour — Indian J. Pure
Appl. Phys. 9, 820 (1971) — B.J. Ansari
and S.L. Srivastava

2,5-Dichlorophenol ($C_6H_3OHCl_2$)

IR Spectrum — Indian J. Pure Appl. Phys.
5, 189 (1967) — S.L. Srivastava
Vibrational Analysis — Indian J. Pure
Appl. Phys. 8, 237 (1970) — S.L.
Srivastava
UV Absorption in Vapour — Indian J. Pure
Appl. Phys. 9, 820 (1971) — B.J.
Ansari and S.L. Srivastava
Vibrational Analysis — Spectrochim. Acta
28A, 33 (1972) — J.H.S. Green, D.J.
Harrison and W. Kynaston

2,6-Dichlorophenol ($C_6H_3OHCl_2$)

IR Spectrum — Indian J. Pure Appl. Phys.
9, 401 (1971) — T.S. Varadarajan and
S. Parthasarthy
Vibrational Analysis — Spectrochim. Acta
28A, 33 (1972) — J.H.S. Green, D.J.
Harrison and W. Kynaston

2,6-Dichlorophenol-OD ($C_6H_3Cl_2OD$)

IR Spectrum — Indian J. Pure Appl. Phys.
9, 401 (1971) — T.S. Varadarajan
and S. Parthasarthy

3,4-Dichlorophenol ($C_6H_3OHCl_2$)

IR Spectrum — Indian J. Pure Appl. Phys.
9, 401 (1971) — T.S. Varadarajan
and S. Parthasarthy
Vibrational Analysis — Spectrochim. Acta
28A, 33 (1972) — J.H.S. Green, D.J.
Harrison and W. Kynaston

3,4-Dichlorophenol-OD ($C_6H_3ODCl_2$)

IR Spectrum — Indian J. Pure Appl. Phys.
9, 401 (1971) — T.S. Varadarajan
and S. Parthasarthy

3,5-Dichlorophenol ($C_6H_3OHCl_2$)

IR Spectrum — Indian J. Pure Appl. Phys.
9, 401 (1971) — T.S. Varadarajan
and S. Parthasarthy
IR Spectrum — Indian J. Pure Appl. Phys.
11, 913 (1973) — N.K. Sanyal and
A.N. Pandey
UV Absorption in Vapour — Indian J. Pure
Appl. Phys. 11, 913 (1973) — N.K.
Sanyal and A.N. Pandey

3,5-Dichlorophenol-OD $(C_6H_3ODCl_2)$

IR Spectrum — Indian J. Pure Appl. Phys.
  9, 401 (1971) — T.S. Varadarajan and
  S. Parthasarthy

2,3-Dichloropyridine $(C_5H_3NCl_2)$

Vibrational Analysis — Spectrochim. Acta
  29A, 1177 (1973) — J.H.S. Green, D.J.
  Harrison and M.R. Kipps

2,5-Dichloropyridine $(C_5H_3NCl_2)$

Vibrational Analysis — Spectrochim. Acta
  29A, 1177 (1973) — J.H.S. Green, D.J.
  Harrison and M.R. Kipps

2,6-Dichloropyridine $(C_5H_3NCl_2)$

Vibrational Analysis — Spectrochim. Acta
  29A, 1177 (1973) — J.H.S. Green, D.J.
  Harrison and M.R. Kipps
IR Spectrum — Indian J. Pure Appl. Phys. 11,
  277 (1973) — R.S. Tripathi
UV Absorption in Vapour — Indian J. Pure
  Appl. Phys. 12, 64 (1974) — R.S.
  Tripathi and B.R. Pandey

3,5-Dichloropyridine $(C_5H_3NCl_2)$

Vibrational Analysis — Spectrochim. Acta
  29A, 1177 (1973) — J.H.S. Green, D.J.
  Harrison and M.R. Kipps
IR Spectrum — Indian J. Pure Appl. Phys.
  11, 277 (1973) — R.S. Tripathi
UV Absorption in Vapour — Indian J. Pure
  Appl. Phys. 12, 64 (1974) — R.S.
  Tripathi and B.R. Pandey

2,4-Dichloropyrimidine $(C_4H_2N_2Cl_2)$

UV Absorption in Vapour — Indian J. Pure
  Appl. Phys. 50, 865 (1976) — N.K. Sanyal
  and S.L. Srivastava

4,6-Dichloropyrimidine $(C_4H_2N_2Cl_2)$

UV Absorption in Vapour — Indian J. Pure
  Appl. Phys. 50, 865 (1976) — N.K.
  Sanyal and S.L. Srivastava

o-Dichlorotetrafluorobenzene
  $(C_6F_4Cl_2)$

Vibrational Analysis — Spectrochim. Acta
  33A, 193 (1977) — J.H.S. Green and
  D.J. Harrison

m-Dichlorotetrafluorobenzene
  $(C_6F_4Cl_2)$

Vibrational Analysis — Spectrochim. Acta
  33A, 193 (1977) — J.H.S. Green and
  D.J. Harrison

p-Dichlorotetrafluorobenzene
  $(C_6F_4Cl_2)$

Vibrational Analysis — Spectrochim. Acta
  33A, 193 (1977) — J.H.S. Green and
  D.J. Harrison

2,3-Dichlorotoluene $(C_6H_3CH_3Cl_2)$

UV Absorption in Vapour — Indian J. Pure
  Appl. Phys. 9, 268 (1971) — T.S.
  Varadarajan and A.K. Kalkar
IR Spectrum — Indian J. Pure Appl. Phys.
  9, 333 (1971) — T.S. Varadarajan and
  A.K. Kalkar

2,4-Dichlorotoluene $(C_6H_3CH_3Cl_2)$

Vibrational Analysis — Indian J. Phys.
  34, 554 (1960) — K.K. Deb and S.B.
  Banerjee
UV Absorption in Vapour — Indian J. Phys.
  35, 628 (1961) — J.K. Roy
Vibrational Analysis — Indian J. Phys.
  36, 59 (1962) — K.K. Deb
Fluorescence — Indian J. Phys. 36, 507
  (1962) — J.K. Roy
UV Absorption in Vapour — Indian J. Pure
  Appl. Phys. 6, 137 (1968) — V.K.
  Mehrotra
IR Spectrum — Indian J. Pure Appl. Phys.
  6, 691 (1968) — V.K. Mehrotra
UV Absorption in Vapour — Curr. Sci. 37,
  134 (1968) — G. Thakur and K.N.
  Upadhya
Vibrational Analysis — Indian J. Pure
  Appl. Phys. 7, 107 (1969) — G. Thakur,
  V.B. Singh and N.L. Singh
Vibrational Analysis — Spectrochim. Acta
  27A, 807 (1971) — J.H.S. Green, D.J.
  Harrison and W. Kynaston

56

2,5-Dichlorotoluene ($C_6H_3CH_3Cl_2$)

UV Absorption in Vapour - Indian J. Pure
    Appl. Phys. 6, 137 (1968) - V.K.
    Mehrotra
IR Spectrum - Indian J. Pure Appl. Phys.
    6, 691 (1968) - V.K. Mehrotra
UV Absorption in Vapour - J. Sci. Res. BHU
    (India) 20, 25 (1969) - R. Amni Amma,
    K.P.R. Nair and D.K. Rai
Vibrational Analysis - Indian J. Pure Appl.
    Phys. 8, 115 (1970) - R. Amni Amma,
    K.P.R. Nair and D.K. Rai
Vibrational Analysis - Spectrochim. Acta
    27A, 807 (1971) - J.H.S. Green, D.J.
    Harrison and W. Kynaston
Vibrational Analysis - Spectrochim. Acta
    28A, 33 (1972) - J.H.S. Green, D.J.
    Harrison and W. Kynaston

2,6-Dichlorotoluene ($C_6H_3CH_3Cl_2$)

UV Absorption in Vapour - Indian J. Pure
    Appl. Phys. 6, 137 (1968) - V.K.
    Mehrotra
IR Spectrum - Indian J. Pure Appl. Phys.
    6, 691 (1968) - V.K. Mehrotra
IR Spectrum - Curr. Sci. 37, 343 (1968) -
    T.S. Varadarajan and A.K. Kalkar
Vibrational Analysis - Spectrochim. Acta
    24A, 1705 (1968) - T. Uno, K. Machida
    and K. Honai
IR Spectrum - Indian J. Pure Appl. Phys.
    9, 333 (1971) - T.S. Varadarajan and
    A.K. Kalkar
Vibrational Analysis - Spectrochim. Acta
    27A, 793 (1971) - J.H.S. Green, D.J.
    Harrison and W. Kynaston
UV Absorption in Different States - Indian
    J. Phys. 51B, 173 (1977) - P.K. Mallick
    and S.B. Banerjee

3,4-Dichlorotoluene ($C_6H_3CH_3Cl_2$)

Vibrational Analysis - Indian J. Phys. 34,
    554 (1960) - K.K. Deb and S.B. Banerjee
UV Absorption in Vapour - Indian J. Phys.
    35, 628 (1961) - J.K. Roy
Vibrational Analysis - Indian J. Phys. 36,
    59 (1962) - K.K. Deb
Fluorescence - Indian J. Phys. 36, 507
    (1962) - J.K. Roy
UV Absorption in Different States - Indian
    J. Phys. 37, 299 (1963) - T.N. Misra
IR Spectrum - Indian J. Pure Appl. Phys.
    9, 333 (1971) - T.S. Varadarajan and
    A.K. Kalkar

Vibrational Analysis - Spectrochim. Acta
    27A, 807 (1971) - J.H.S. Green, D.J.
    Harrison and W. Kynaston

3,6-Dichlorotoluene ($C_6H_3CH_3Cl_2$)

Vibrational Analysis - Spectrochim. Acta
    27A, 807 (1971) - J.H.S. Green, D.J.
    Harrison and W. Kynaston

3,5-Dichlorotrifluoropyridine
    ($C_5NCl_2F_3$)

Vibrational Analysis - Spectrochim. Acta
    33A, 81 (1977) - J.H.S. Green and
    D.J. Harrison

o-Dicyanobenzene ($C_6H_4(CN)_2$)

UV Absorption Spectrum - Bull. Chem. Soc.
    Jap. 48, 420 (1975) - H. Morita, S.
    Matsumoto and S. Nagakura

m-Dicyanobenzene ($C_6H_4(CN)_2$)

UV Absorption Spectrum - Bull. Chem. Soc.
    Jap. 48, 420 (1975) - H. Morita, S.
    Matsumoto and S. Nagakura

p-Dicyanobenzene ($C_6H_4(CN)_2$)

UV Absorption Spectrum - Bull. Chem. Soc.
    Jap. 48, 420 (1975) - H. Morita, S.
    Matsumoto and S. Nagakura

N,N-Diethylaniline ($C_6H_5N(C_2H_5)_2$)

Raman - Monatsh. Chem. 69, 363 (1936) -
    L. Kahovec and A.W. Reitz
Vibrational Analysis - Indian J. Pure
    Appl. Phys. 9, 336 (1971) - V.N. Verma,
    K.P.R. Nair and D.K. Rai
UV Absorption Spectrum - Chem. Phys. Lett.
    21, 511 (1973) - H. Tsubomura and T.
    Sakata
UV Absorption Spectrum - Chem. Phys. Lett.
    28, 39 (1974) - A. Davidson and B.
    Norden
UV Absorption Spectrum - J. Mol. Spectrosc.
    64, 139 (1977) - K. Fuke and S. Naga-
    kura

o-Diethylbenzene ($C_6H_4(C_2H_5)_2$)

Vibrational Analysis — Anal. Chem. 19, 700
    (1947) — M.R. Fenske, W.G. Braun, R.V.
    Wiegand, D. Quiggle, R.M. McCormick
    and D.H. Rank

m-Diethylbenzene ($C_6H_4(C_2H_5)_2$)

Vibrational Analysis — Anal. Chem. 19, 700
    (1947) — M.R. Fenske, W.G. Braun, R.V.
    Wiegand, D. Quiggle, R.M. McCormick
    and D.H. Rank
Vibrational Analysis — Izv. Akad. Nauk.
    Otdel Ser Khim. 1437 (1961) — V.T.
    Aleksanian, Kh.E. Sterin, S.A. Ukhelin,
    O.V. Bragain, E.A. Mikhaileva, E.N.
    Smirnova, N.I. Tiunkina and B.A.
    Kazanski

p-Diethylbenzene ($C_6H_4(C_2H_5)_2$)

Vibrational Analysis — Izv. Akad. Nauk.
    Otdel. Ser. Khim. 1437 (1961) — V.T.
    Aleksanian, Kh.E. Sterin, S.A. Ukhelin,
    O.V. Bragain, E.A. Mikhaileva, E.N.
    Smirnova, N.I. Tiunkina and B.A.
    Kazanski

N,N-Diethyl-p-Nitroaniline ($C_6H_4NO_2N(C_2H_5)_2$)

Vibrational Analysis — Z. Elektrochem. 64,
    853 (1960) — H.W. Schrotter
Vibrational Analysis — Spectrochim. Acta
    17, 523 (1961) — J. Brandmuller, E.W.
    Schmid, H.W. Schrotter and G. Nonnen-
    macher

2,3-Diethyl-p-Xylene ($C_6H_2(C_2H_5)_2(CH_3)_2$)

Vibrational Analysis — Vestnik. Moskov Univ.
    17, 66 (1962) — E.G. Treshova, V.R.
    Skvartchenke and R.I. Levina

2,3-Difluoroaniline ($C_6H_3NH_2F_2$)

IR Spectrum — Curr. Sci. 40, 655 (1971) —
    B.B. Lal, G.D. Baruah and I.S. Singh

2,5-Difluoroaniline ($C_6H_3NH_2F_2$)

Vibrational Analysis — Proc. Roy. Soc.
    A243, 143 (1957) — P.J. Krueger and
    H.W. Thompson
Vibrational Analysis — Can. J. Chem. 40,
    2300 (1962) — P.J. Krueger
IR Spectrum — Aust. J. Chem. 15, 626
    (1962) — A.N. Hamby and B.V.O. Grady
UV Absorption in Vapour — Indian J. Pure
    Appl. Phys. 3, 497 (1963) — N.L.
    Singh and S.N. Singh
IR Spectrum — Indian J. Pure Appl. Phys.
    7, 250 (1969) — S.N. Singh and N.L.
    Singh

o-Difluorobenzene ($C_6H_4F_2$)

UV Absorption in Vapour — J. Chem. Phys.
    22, 234 (1954) — C.D. Cooper
Vibrational Analysis — Bull. Amer. Phys.
    Soc. 2(2), 99 (1956) — C.J. Halley
Vibrational Analysis — Spectrochim. Acta
    18, 915 (1962) — D. Steele
Vibrational Analysis — J. Chem. Soc. 473
    (1963) — J.H.S. Green, W. Kynaston
    and H.M. Paisley
Vibrational Analysis — J. Chem. Phys. 38,
    532 (1963) — D.W. Scott, J.F. Messerly,
    S.S. Todd, I.A. Hossenlopp, A. Osborn
    and J.P. McCullough
UV Absorption in Solution — J. Chim. Phys.
    62, 347 (1965) — M. Brigodiot and
    J.M. Lebas
IR Spectrum — Curr. Sci. 36, 231 (1967) —
    A.N. Pathak and L.N. Tripathi
Fluorescence — Can. J. Chem. 46, 3177
    (1968) — J.L. Durham, G.P. Seneluk
    and I. Unger
Vibrational Analysis — Spectrochim. Acta
    26A, 1913 (1970) — J.H.S. Green
UV Emission in Vapour — Indian J. Pure
    Appl. Phys. 9, 815 (1971) — J.V.
    Shukla, K.N. Upadhya and D.K. Rai
Fluorescence — Chem. Phys. Lett. 14,
    404 (1972) — G.M. Breuer and E.K.C.
    Lee
Calculation — J. Mol. Spectrosc. 48, 446
    (1973) — V.J. Eaton and D. Steele
Calculation — J. Mol. Struct. 32, 93
    (1976) — F. Torok, A. Hegedus, K.
    Kasa and P. Pulay
Vibrational Analysis — J. Mol. Spectrosc.
    64, 1 (1977) — B. Lunelli and M.G.
    Giorgini

o—Difluorobenzene—D_4 ($C_6D_4F_2$)

Vibrational Analysis — J. Mol. Spectrosc.
64, 1 (1977) — B. Lunelli and M.G.
Giorgini

m—Difluorobenzene ($C_6H_4F_2$)

Raman — Monatsh. Chem. 74, 160 (1943) —
E. Herz

UV Absorption in Vapour — Phys. Rev. 87,
213 (1952) — V.R.K.S. Rao and H. Sponer

Vibrational Analysis — J. Chem. Phys. 21,
1470 (1953) — E.E. Ferguson, R.L.
Collins, J.R. Nielsen and D.C. Smith

IR Spectrum — Spectrochim. Acta 7, 253
(1955) — D.H. Whiffen

Vibrational Analysis — J. Chem. Soc. 473
(1963) — J.H.S. Green, W. Kynaston and
H.M. Paisley

UV Emission in Vapour — Indian J. Phys. 42,
42 (1968) — S.N. Singh and I.S. Singh

UV Absorption in Vapour — Indian J. Pure
Appl. Phys. 7, 39 (1969) — P.D. Singh
and A.N. Pathak

Vibrational Analysis — Spectrochim. Acta
26A, 1523 (1970) — J.H.S. Green

UV Emission in Vapour — Indian J. Pure
Appl. Phys. 9, 815 (1971) — J.V. Shukla,
K.N. Upadhya and D.K. Rai

Fluorescence — J. Phys. Chem. 75, 1233
(1971) — T.L. Brewer

Fluorescence — Chem. Phys. Lett. 14, 404
(1972) — G.M. Breuer and E.K.C. Lee

Calculation — J. Mol. Spectrosc. 48, 446
(1973) — V.J. Eaton and D. Steele

Calculation — J. Mol. Struct. 32, 93
(1976) — F. Torok, A. Hegedus, K. Kosa
and P.Pulay

p—Difluorobenzene ($C_6H_4F_2$)

Vibrational Analysis — J. Chem. Phys. 18,
1680 (1950) — A.H. Delsemme

Vibrational Analysis — J. Chem. Phys. 21,
1457, 1727 (1953) — E.E. Ferguson,
R.L. Hudson, J.R. Nielsen and D.C.
Smith

Vibrational Analysis — J. Chem. Phys. 21,
1736 (1953) — E.E. Ferguson, R.L.
Hudson, J.R. Nielsen and D.C. Smith

Vibrational Analysis — Spectrochim. Acta
6, 225 (1954) — J.M. Lebas, C.
Garrigou-Lagrange and M.L. Josien

UV Absorption in Vapour — J. Chem. Phys.
22, 503 (1954) — C.D. Cooper

UV Absorption in Vapour — J. Chem. Phys.
23, 796 (1955) — T. Anno and I.
Matubara

Vibrational Analysis — Spectrochim. Acta
12, 47 (1958) — A. Stojilkovic and
D.H. Whiffen

Vibrational Analysis — J. Chem. Soc. 473
(1963) — J.H.S. Green, W. Kynaston
and H.M. Paisley

IR Spectrum — Spectrochim. Acta 19, 785
(1963) — D. Steele, W. Kynaston and
H.A. Gebbie

UV Absorption in Vapour — Philos. Trans.
R. Soc. Lond. A259, 499 (1966) —
J.H. Colloman, T.M. Dunn and I.M.
Mills

Vibrational Analysis — Proc. Roy. Soc.
A298, 51 (1967) — P.R. Griffiths and
H.W. Thompson

Vibrational Analysis — J. Chim. Phys. 64,
1450 (1967) — P. Pommez, M. Lafaix,
P. Delorme and V. Lorenzelli

Fluorescence — Can. J. Chem. 46, 3177
(1968) — J.L. Durham, G.P. Semeluk
and I. Unger

UV Absorption in Vapour (High Resolution)—
Indian J. Pure Appl. Phys. 7, 765
(1969) — S.N. Thakur and N.L. Singh

UV Absorption in Vapour — J. Mol. Spectrosc. 36, 541 (1970) — Y. Udagawa, M.
Ito and S. Nagakura

UV Absorption in Vapour — Mol. Phys. 18,
783 (1970) — T. Cvitas and J.M. Hollas

Fluorescence — Curr. Sci. 37, 344 (1970) —
J.V. Shukla, M.M. Rai and K.N. Upadhya

Vibrational Analysis — Spectrochim. Acta
26A, 1503 (1970) — J.H.S. Green

UV Emission in Vapour — Indian J. Pure
Appl. Phys. 9, 815 (1971) — J.V.
Shukla, K.N. Upadhya and D.K. Rai

UV Absorption in Vapour — J. Mol.
Spectrosc. 40, 262 (1971) — A. Hartford, Jr and J.R. Lombardi

Fluorescence — Chem. Phys. Lett. 14, 404
(1972) — G.M. Breuer and E.K.C. Lee

Calculation — Chem. Phys. Lett. 19, 445
(1973) — J.S. Yadav, P.C. Mishra
and D.K. Rai

Calculation — J. Mol. Spectrosc. 48,
446 (1973) — V.J. Eaton and D. Steele

Calculation — J. Mol. Struct. 32, 93
(1976)— F. Torok, A. Hegedus, K. Kosa
and P. Pulay

Fluorescence — J. Chem. Phys. 67, 236,
242 (1977) — L.J. Volk and E.K.C. Lee

p-Difluorobenzene-D$_4$ (C$_6$D$_4$F$_2$)

IR Spectrum — Spectrochim. Acta 19, 785
(1963) — D. Steele, W. Kynaston and
H.A. Gebbie
Vibrational Analysis — Spectrochim. Acta
25A, 507 (1969) — P.N. Gates, K.
Radcliffe and D. Steele

2,5-Difluorobenzotrifluoride
(C$_6$H$_3$F$_2$CF$_3$)

Vibrational Analysis — J. Chem. Soc. 1432
(1948) — H.W. Thompson and N.N. Temple

2,4-Difluoronitrobenzene (C$_6$H$_3$F$_2$NO$_2$)

Vibrational Analysis — Indian J. Phys. 48,
53 (1974) — J.V. Shukla, K.N. Upadhya
and D.K. Rai

2,5-Difluoronitrobenzene (C$_6$H$_3$F$_2$NO$_2$)

Vibrational Analysis — Indian J. Phys. 48,
53 (1974) — J.V. Shukla, K.N. Upadhya
and D.K. Rai

2,6-Difluoropyridine (C$_5$H$_3$NF$_2$)

Vibrational Analysis — Spectrochim. Acta
23A, 2997 (1967) — R.T. Bailey and D.
Steele
Vibrational Analysis — Spectrochim. Acta
29A, 1177 (1973) — J.H.S. Green, D.J.
Harrison and M.R. Kipps
Vibrational Analysis — Spectrochim. Acta
34A, 583 (1978) — S. Lui, S. Suzuki
and J.A. Ladd

2,6-Difluoro-4-deuteropyridine
(C$_5$H$_2$DF$_2$)

Vibrational Analysis — Spectrochim. Acta
34A, 583 (1978) — S. Lui, S. Suzuki
and J.A. Ladd

2,6-Difluoro-3,4,5-trideuteropyridine
(C$_5$D$_3$F$_2$)

Vibrational Analysis — Spectrochim. Acta
34A, 583 (1978) — S. Lui, S. Suzuki
and J.A. Ladd

2,4-Difluorotoluene (C$_6$H$_3$CH$_3$F$_2$)

Vibrational Analysis — J. Chem. Soc. 1432
(1948) — H.W. Thompson and N.N. Temple
IR Spectrum — Indian J. Pure Appl. Phys.
7, 107 (1969) — G. Thakur, V.B. Singh
and N.L. Singh
Vibrational Analysis — Spectrochim. Acta
27A, 807 (1971) — J.H.S. Green, D.J.
Harrison and W. Kynaston

2,5-Difluorotoluene (C$_6$H$_3$CH$_3$F$_2$)

Vibrational Analysis — Spectrochim. Acta
27A, 807 (1971) — J.H.S. Green, D.J.
Harrison and W. Kynaston

3,6-Difluorotoluene (C$_6$H$_3$CH$_3$F$_2$)

Vibrational Analysis — Spectrochim. Acta
27A, 807 (1971) — J.H.S. Green, D.J.
Harrison and W. Kynaston

o-Dihydroxybenzene (C$_6$H$_4$(OH)$_2$)

Vibrational Analysis — Spectrochim. Acta
16, 528 (1960) — A. Hidalgo and C.
Otero
Vibrational Analysis — Spectrochim. Acta
17, 1049 (1961) — G. Nonnenmacher and
R. Mecke
Vibrational Analysis — Appl. Spectrosc.
16, 32 (1962) — R.J. Jackobsen and
E.J. Brewer
Calculation — J. Mol. Struct. 6, 246 (1970)-
P.C. Mishra and D.K. Rai
Calculation — J. Mol. Struct. 13, 253
(1972) — J.S. Yadav, P.C. Mishra and
D.K. Rai
Vibrational Analysis — Spectrochim. Acta
30A, 2141 (1974) — H.W. Wilson

m-Dihydroxybenzene (C$_6$H$_4$(OH)$_2$)

UV Absorption in Vapour — J. Chem. Phys.
18, 1135 (1950) — C.A. Beck
Vibrational Analysis — Spectrochim. Acta
16, 528 (1960) — A. Hidalgo and C.
Otero
Vibrational Analysis — Spectrochim. Acta
17, 1049 (1961) — G. Nonnenmacher
and R. Mecke
Vibrational Analysis — Appl. Spectrosc.
16, 32 (1962) — R.J. Jackobsen and
E.J. Brewer

Calculation — J. Mol. Struct. <u>6</u>, 246
    (1970) — P.C. Mishra and D.K. Rai
Calculation — J. Mol. Struct. <u>13</u>, 253
    (1972) — J.S. Yadav, P.C. Mishra and
    D.K. Rai
Vibrational Analysis — Spectrochim. Acta
    <u>30A</u>, 2141 (1974) — H.W. Wilson

p—Dihydroxybenzene $(C_6H_4(OH)_2)$

UV Absorption in Vapour — J. Chem. Phys.
    <u>18</u>, 1135 (1950) — C.A. Beck
Vibrational Analysis — Spectrochim. Acta
    <u>16</u>, 528 (1960) — A. Hidalgo and C.
    Otero
Vibrational Analysis — Spectrochim. Acta
    <u>17</u>, 1049 (1961) — G. Nonnenmacher and
    R. Mecke
Vibrational Analysis — Appl. Spectrosc.
    <u>16</u>, 32 (1962) — R.J. Jackobsen and
    E.J. Brewer
Calculation — J. Mol. Struct. <u>6</u>, 246
    (1970) — P.C. Mishra and D.K. Rai
Calculation — J. Mol. Struct. <u>13</u>, 253
    (1972) — J.S. Yadav, P.C. Mishra and
    D.K. Rai
Vibrational Analysis — Spectrochim. Acta
    <u>30A</u>, 2141 (1974) — H.W. Wilson

2,5—Dihydroxy—p—benzoquinone $(C_6H_2O_2(OH)_2)$

UV Absorption in Solution — Z. Naturforsch.
    <u>10B</u>, 668 (1955) — W. Fraig, Th. Plvetz
    and A. Kulhner
UV Absorption in Vapour — Indian J. Pure
    Appl. Phys. <u>5</u>, 572 (1967) — S.C.
    Srivastava and R.S. Singh
IR Spectrum — Indian J. Pure Appl. Phys.
    <u>5</u>, 572 (1967) — S.C. Srivastava and
    R.S. Singh

1,4—Dihydroxy—2—Methylanthraquinone $(C_{15}H_{10}O_4)$

IR Spectrum — Curr. Sci. <u>41</u>, 665 (1972) —
    S. Nath Singh, G.D. Baruah and R.S.
    Singh

1,8—Dihydroxy—3—Methylanthraquinone $(C_{15}H_{10}O_4)$

IR Spectrum — Curr. Sci. <u>41</u>, 665 (1972) —
    S. Nath Singh, G.D. Baruah and R.S.
    Singh

2,6—Dihydroxynaphthol $(C_{10}H_6(OH)_2)$

Vibrational Analysis — Indian J. Phys.
    <u>51B</u>, 93 (1977) — O.P. Sharma and R.D.
    Singh

o—Diiodobenzene $(C_6H_4I_2)$

Vibrational Analysis — J. Chim. Phys.
    <u>62</u>, 347 (1965) — M. Brigodiot and J.M.
    Lebas
Vibrational Analysis — Spectrochim. Acta
    <u>26A</u>, 1913 (1970) — J.H.S. Green

m—Diiodobenzene $(C_6H_4I_2)$

Vibrational Analysis — Spectrochim. Acta
    <u>26A</u>, 1523 (1970) — J.H.S. Green

p—Diiodobenzene $(C_6H_4I_2)$

Raman — Spectrochim. Acta <u>12</u>, 47 (1958) —
    A. Stojilkovic and D.H. Whiffen
Vibrational Analysis — Spectrochim. Acta
    <u>12</u>, 305 (1958) — C. Garrigou-Lagrange,
    J.M. Lebas and M.L. Josien
Vibrational Analysis — Spectrochim. Acta
    <u>13</u>, 225 (1959) — J.M. Lebas, C.
    Garrigou-Lagrange and M.L. Josien
UV Absorption in Solution — Can. J. Chem.
    <u>39</u>, 2295 (1961) — W.B. Forbes
IR Spectrum — Proc. Roy. Soc. <u>A298</u>, 51
    (1967) — P.R. Griffiths and H.W.
    Thompson
Vibrational Analysis — J. Chim. Phys. <u>64</u>,
    1450 (1967) — P. Pommez, M. Lafaix,
    P. Delorme and V. Lorenzelli
Vibrational Analysis — Spectrochim. Acta
    <u>26A</u>, 1503 (1970) — J.H.S. Green

o—Diiodotetrafluorobenzene $(C_6I_2F_4)$

Vibrational Analysis — Spectrochim. Acta
    <u>33A</u>, 193 (1977) — J.H.S. Green and
    D.J. Harrison

m—Diiodotetrafluorobenzene $(C_6I_2F_4)$

Vibrational Analysis — Spectrochim. Acta
    <u>33A</u>, 193 (1977) — J.H.S. Green and
    D.J. Harrison

p—Diiodotetrafluorobenzene $(C_6I_2F_4)$

Vibrational Analysis — Spectrochim. Acta
33A, 193 (1977) — J.H.S. Green and D.J.
Harrison

2,5—Dimethoxyaniline $(C_6H_3NH_2(OCH_3)_2)$

Vibrational Analysis — J. Sci. Res. BHU
(India) 28(2), 357 (1977) — V.N. Verma

2,3—Dimethoxybenzaldehyde
$(C_6H_3CHO(OCH_3)_2)$

UV Absorption in Solution — Indian J. Phys.
49, 494 (1975) — S.C. Srivastava, B.B.
Lal and I.S. Singh
IR Spectrum — Indian J. Pure Appl. Phys. 16,
939 (1978) — S.J. Singh and R. Singh

2,4—Dimethoxybenzaldehyde
$(C_6H_3CHO(OCH_3)_2)$

UV Ansorption in Solution — Indian J. Phys.
49, 494 (1975) — S.C. Srivastava, B.B.
Lal and I.S. Singh
IR Spectrum — Indian J. Pure Appl. Phys.
16, 939 (1978) — S.J. Singh and R. Singh

3,4—Dimethoxybenzaldehyde
$(C_6H_3CHO(OCH_3)_2)$

UV Absorption in Solution — Indian J. Phys.
49, 494 (1975) — S.C. Srivastava, B.B.
Lal and I.S. Singh
IR Spectrum — Indian J. Pure Appl. Phys.
16, 939 (1978) — S.J. Singh and R. Singh

o—Dimethoxybenzene $(C_6H_4(OCH_3)_2)$

Raman — Monatsh. Chem. 64, 374 (1934) —
K.W.F. Kohlrausch and A. Pongratz
Raman — Sci. Pap. Inst. Phys. Chem. Res.
(Tokyo) 34, 1147 (1938) — S. Mizushima,
Y. Morino and H. Okazuki
IR Spectrum — J. Phys. Radium 9, 13 (1938) —
J. Lecomte
Raman — Monatsh. Chem. 76, 1 (1946) — E.
Herz
IR Spectrum — J. Chem. Soc. 3670 (1959) —
A.R. Katritzky and R.A. Jones
UV Absorption in Different States — Indian
J. Phys. 46, 49 (1972) — D. Marjit, P.K.
Bishnui and S.B. Banerjee

Vibrational Analysis — Indian J. Phys.
46, 49 (1972) — D. Marjit, P.K.
Bishnui and S.B. Banerjee
Emission Spectrum (Electron Impact) — Bull.
Chem. Soc. Jap. 48, 645 (1975) — T.
Ogawa, T. Isnasaka, M. Toyoda, M.
Tsuji and N. Ishibashi

m—Dimethoxybenzene $(C_6H_4(OCH_3)_2)$

Raman — Monatsh. Chem. 64, 361 (1934) —
K.W.F. Kohlrausch and A. Pongratz
Raman — Monatsh. Chem. 64, 374 (1934) —
K.W.F. Kohlrausch and A. Pongratz
Raman — Sci. Pap. Inst. Phys. Chem. Res.
(Tokyo) 34, 1147 (1938) — S. Mizushima,
Y. Morino and H. Okazuki
IR Spectrum — J. Phys. Radium 9, 13
(1938) — J. Lecomte
Raman — Monatsh. Chem. 74, 160 (1941) —
E. Herz
Raman — Monatsh. Chem. 76, 1 (1946) — E.
Herz
UV Absorption in Vapour — Indian J. Phys.
33, 241 (1959) — K. Sreeramamurty
IR Spectrum — J. Chem. Soc. 2051 (1959) —
A.R. Katritzky and P. Simons
IR Spectrum — J. Chem. Soc. 2058 (1959) —
A.R. Katritzky and P. Simons
UV Absorption in Different States —
Indian J. Phys. 46, 457 (1972) — D.
Marjit, P.K. Bishnui and S.B. Banerjee
Vibrational Analysis — Indian J. Phys. 46,
457 (1972) — D. Marjit, P.K. Bishnui
and S.B. Banerjee
Emission Spectrum (Electron Impact) — Bull.
Chem. Soc. Jap. 48, 645 (1975) — T.
Ogawa, T. Imasaka, M. Toyoda, M. Tsuji
and N. Ishibashi

p—Dimethoxybenzene $(C_6H_4(OCH_3)_2)$

Raman — Monatsh. Chem. 64, 361 (1934) —
K.W.F. Kohlrausch and A. Pongratz
Raman — Monatsh. Chem. 64, 374 (1934) —
K.W.F. Kohlrausch and A. Pongratz
Raman — Sci. Pap. Inst. Phys. Chem. Res.
(Tokyo) 34, 1147 (1938) — S. Mizushima,
Y. Morino and H. Okazuki
IR Spectrum — J. Phys. Radium 9, 13 (1938) —
J. Lecomte
Raman — Monatsh. Chem. 76, 1 (1946) — E.
Herz
Raman — Indian J. Phys. 31, 99 (1957) —
S.K. Sen
UV Absorption in Different States — Indian
J. Phys. 31, 99 (1957) — S.K. Sen

Vibrational Analysis — Spectrochim. Acta
12, 305 (1958) — C. Garrigou-Lagrange,
J.M. Lebas and M.L. Josien
IR Spectrum — J. Chem. Soc. 2051 (1959) —
A.R. Katritzky and P. Simons
IR Spectrum — J. Chem. Soc. 2058 (1959) —
A.R. Katritzky and P. Simons
Emission Spectrum in Vapour — Indian J. Pure
Appl. Phys. 5, 278 (1967) — S.C.
Srivastava and R.S. Singh
UV Absorption in Vapour — Indian J. Pure
Appl. Phys. 5, 278 (1967) — S.C.
Srivastava and R.S. Singh
Vibrational Analysis — Indian J. Phys. 46,
457 (1972) — D. Marjit, P.K. Bishnui
and S.B. Banerjee
UV Absorption in Different States — Indian
J. Phys. 46, 457 (1972) — D. Marjit,
P.K. Bishnui and S.B. Banerjee
Emission Spectrum (Electron Impact) — Bull.
Chem. Soc. Jap. 48, 645 (1975) — T.
Ogawa, T. Imasaka, M. Toyoda, M. Tzuji
and N. Ishibashi

2,4-Dimethoxypyridine $(C_5H_3N(OCH_3)_2)$

UV Absorption in Solution — Spectrochim.
Acta 24A, 207 (1968) — G. Favini, M.
Raimondo and G. Gondolfo

2,6-Dimethoxypyridine $(C_5H_3N(OCH_3)_2)$

UV Absorption in Solution — Spectrochim.
Acta 24A, 207 (1968) — G. Favini, M.
Raimondo and G. Gondolfo

p-Dimethylaminoazobenzene $(C_{14}H_{15}N_3)$

Raman (Resonance) — Spectrochim. Acta 32A,
1179 (1976) — T. Uno, B-K. Kim, Y.
Saito and K. Machida

p-Dimethylaminoazobenzene-D$_5$
$(C_{14}H_{10}D_5N_3)$

Raman (Resonance) — Spectrochim. Acta 32A,
1179 (1976) — T. Uno, B-K. Kim, Y.
Saito and K. Machida

N,N-Dimethylaniline $(C_6H_5N(CH_3)_2)$

Raman — Monatsh. Chem. 69, 363 (1936) —
L. Kahovec and A.W. Reitz

Vibrational Analysis — Indian J. Pure
Appl. Phys. 9, 336 (1971) — V.N.
Verma, K.P.R. Nair and D.K. Rai
Vibrational Analysis — Indian J. Pure
Appl. Phys. 16, 454 (1978) — A.K.
Ansari and P.K. Verma

2,3-Dimethylaniline $(C_6H_3NH_2(CH_3)_2)$

IR Spectrum — Indian J. Pure Appl. Phys.
13, 718 (1975) — M. Prasad

2,4-Dimethylaniline $(C_6H_3NH_2(CH_3)_2)$

IR Spectrum — Indian J. Pure Appl. Phys.
13, 718 (1975) — M. Prasad

3,4-Dimethylaniline $(C_6H_3NH_2(CH_3)_2)$

IR Spectrum — Indian J. Pure Appl. Phys.
13, 718 (1975) — M. Prasad

3,5-Dimethylaniline $(C_6H_3NH_2(CH_3)_2)$

Vibrational Analysis — Spectrochim. Acta
27A, 807 (1971) — J.H.S. Green, D.J.
Harrison and W. Kynaston

Dimethylbenzene $(C_6H_4(CH_3)_2)$

See   Xylene

N,N-Dimethyl-p-Nitroaniline
$(C_6H_4NO_2N(CH_3)_2)$

Vibrational Analysis — Z. Elektrochem.
64, 853 (1960) — H.W. Schrotter
Vibrational Analysis — Spectrochim. Acta
17, 523 (1961) — J. Brandmuller, E.W.
Schmid, H.W. Schrotter and G. Nonnenmacher

Dimethylphenol $(C_6H_3OH(CH_3)_2)$

See   Xylenol

2,3-Dimethylpyridine $(C_5H_3N(CH_3)_2)$

IR Spectrum — J. Chem. Soc. 1934 (1959) —
E.A. Coulson, J.D. Con, E.F. Herrington and J.F. Mortin

Vibrational Analysis — Spectrochim. Acta
26A, 2139 (1970) — J.H.S. Green, D.J.
Harrison, W. Kynaston and H.M. Paisley
Vibrational Analysis — Indian J. Phys.
46, 300 (1972) — K.C. Medhi

2,4-Dimethylpyridine ($C_5H_3N(CH_3)_2$)

UV Absorption in Vapour — J. Chem. Soc.
97, 692 (1910) — J.E. Purvis
Raman — Boll. Sci. Fac. Chim. Ind. Bologna
1, 137 (1940) — R. Manzani-Ansidei
Raman — Boll. Sci. Fac. Chim. Ind. Bologna
1, 184 (1940) — R. Manzani-Ansidei
Raman — Z. Phys. Chem. 53B, 124 (1943) —
E. Herz, L. Kahovec and K.W.F. Kohl-
rausch
Vibrational Analysis — Spectrochim. Acta
21, 895 (1965) — K.C. Medhi and D.K.
Mukharjee
Vibrational Analysis — Spectrochim. Acta
26A, 2139 (1970) — J.H.S. Green, D.J.
Harrison, W. Kynaston and H.M. Paisley
UV Absorption in Different States — Indian
J. Phys. 45, 155 (1971) — S.C. Bag

2,5-Dimethylpyridine ($C_5H_3N(CH_3)_2$)

Vibrational Analysis — Spectrochim. Acta
26A, 2139 (1970) — J.H.S. Green, D.J.
Harrison, W. Kynaston and H.M. Paisley
Vibrational Analysis — Spectrochim. Acta
29A, 1177 (1973) — J.H.S. Green, D.J.
Harrison and M.R. Kipps

2,6-Dimethylpyridine ($C_5H_3N(CH_3)_2$)

Raman — Boll. Sci. Fac. Chim. Ind. Bologna
1, 137 (1940) — R. Manzani-Ansidei
Raman — Boll. Sci. Fac. Chim. Ind. Bologna
1, 184 (1940) — R. Manzani-Ansidei
Raman — Z. Phys. Chem. 53B, 124 (1945) —
E. Herz, L. Kahovec and K.W.F. Kohl-
rausch
UV Absorption in Vapour — Disc. Faraday
Soc. 9, 26 (1950) — E.F. Herrington
Vibrational Analysis — Spectrochim. Acta
21, 895 (1965) — K.C. Medhi and D.K.
Mukharjee
Vibrational Analysis — Spectrochim. Acta
21, 105 (1965) — M. Goldestein, E.F.
Mooney, A. Anderson and H.A. Gebbie
Vibrational Analysis — Spectrochim. Acta
23A, 2997 (1967) — R.T. Bailey and
D. Steele

Phosphorescence — J. Amer. Chem. Soc. 91,
6508 (1969) — R.T. Hoover and M. Kasha
Vibrational Analysis — Spectrochim. Acta
26A, 2139 (1970) — J.H.S. Green, D.J.
Harrison, W. Kynaston and H.M. Paisley
UV Absorption in Different States — Indian
J. Phys. 45, 115 (1971) — S.C. Bag
Vibrational Analysis — Spectrochim. Acta
29A, 1177 (1973) — J.H.S. Green, D.J.
Harrison and M.R. Kipps

3,4-Dimethylpyridine ($C_5H_3N(CH_3)_2$)

Vibrational Analysis — Spectrochim. Acta
26A, 2139 (1970) — J.H.S. Green, D.J.
Harrison, W. Kynaston and H.M. Paisley

3,5-Dimethylpyridine ($C_5H_3N(CH_3)_2$)

Vibrational Analysis — Spectrochim. Acta
26A, 2139 (1970) — J.H.S. Green, D.J.
Harrison , W. Kynaston and H.M. Paisley

2,3-Dinitroaniline ($C_6H_3NH_2(NO_2)_2$)

IR Spectrum — Aust. J. Chem. 11, 513
(1958) — L.K. Dyall and A.N. Hamby

2,4-Dinitroaniline ($C_6H_3NH_2(NO_2)_2$)

IR Spectrum — Aust. J. Chem. 11, 513
(1958) — L.K. Dyall and A.N. Hamby
IR Spectrum — Spectrochim. Acta 17, 291
(1961) — L.K. Dyall
IR Spectrum — Aust. J. Chem. 23, 947
(1970) — L.K. Dyall

2,6-Dinitroaniline ($C_6H_3NH_2(NO_2)_2$)

IR Spectrum — Aust. J. Chem. 11, 513
(1958) — L.K. Dyall and A.N. Hamby
IR Spectrum — Spectrochim. Acta 17, 291
(1961) — L.K. Dyall

o-Dinitrobenzene ($C_6H_4(NO_2)_2$)

Raman — Izv. Akad. Nauk. SSSR 12, 553
(1948) — Ya.S. Bobovich and M.V.
Vol'Kenshtein
IR Spectrum — J. Chem. Soc. 2051 (1959) —
A.R. Katritzky and P. Simons
IR Spectrum — J. Chem. Soc. 3273 (1959) —
C.P. Conduit

64

IR Spectrum — Anal. Chem. <u>32</u>, 495 (1960) —
F. Pristera, M. Halik, A. Castelli and
W. Fredericks
Vibrational Analysis — J. Chim. Phys. <u>62</u>,
347 (1965) — M. Brigodiot and J.M.
Lebas
Vibrational Analysis — Indian J. Phys. <u>42</u>,
511 (1968) — J.V. Shukla, V.B. Singh
and K.N. Upadhya
Raman — Indian J. Pure Appl. Phys. <u>7</u>, 830
(1969) — J.V. Shukla and K.N. Upadhya
Vibrational Analysis — Spectrochim. Acta
<u>27A</u>, 817 (1971) — J.H.S. Green and
H.A. Lauwers
Calculation — J. Mol. Struct. <u>18</u>, 457
(1973) — A.I. Kiss and J. Szoke

m-Dinitrobenzene $(C_6H_4(NO_2)_2)$

Raman — Izv. Akad. Nauk. SSSR <u>12</u>, 553
(1948) — Ya.S. Bobovich and M.V. Vol-
Kenshtein
IR Spectrum — J. Amer. Chem. Soc. <u>77</u>,
6341 (1955) — J.F. Brown, Jr
IR Spectrum — J. Chem. Soc. 2051 (1959) —
A.R. Katritzky and P. Simons
IR Spectrum — J. Chem. Soc. 3273 (1959) —
C.P. Conduit
Vibrational Analysis — Spectrochim. Acta
<u>16</u>, 106 (1960) — N. Fuson, C. Garrigou-
Lagrange and M.L. Josien
IR Spectrum — Anal. Chem. <u>32</u>, 495 (1960) —
F. Pristera, M. Holik, A. Castelli
and W. Fredericks
Raman — Opt. Spektrosk. <u>8</u>, 22 (1960) — Ya.
S. Bobovich and M. Ya. Tsentev
Vibrational Analysis — J. Chim. Phys. <u>63</u>,
552 (1966) — C. Garrigou-Lagrange, M.
Chehata and J. Lascombe
Vibrational Analysis — Indian J. Phys. <u>42</u>,
511 (1968) — J.V. Shukla, V.B. Singh
and K.N. Upadhya
Raman — Indian J. Pure Appl. Phys. <u>7</u>, 830
(1969) — J.V. Shukla and K.N. Upadhya
Vibrational Analysis — Spectrochim. Acta
<u>27A</u>, 817 (1971) — J.H.S. Green and
H.A. Lauwers
Calculation — J. Mol. Struct. <u>18</u>, 457
(1973) — A.I. Kiss and J. Szoke
UV Absorption in Vapour — Spectrochim. Acta
<u>33A</u>, 21 (1977) — S. Millefieri, G.
Favini, A. Millefieri and D. Grasso

p-Dinitrobenzene $(C_6H_4(NO_2)_2)$

IR Spectrum — J. Amer. Chem. Soc. <u>78</u>, 4225
(1956) — R.D. Kross and V.A. Fassel

Vibrational Analysis — Spectrochim. Acta
<u>12</u>, 305 (1958) — C. Garrigou-Lagrange,
J.M. Lebas and M.L. Josien
IR Spectrum — J. Chem. Soc. 2051 (1959) —
A.R. Katritzky and P. Simons
IR Spectrum — J. Chem. Soc. 3273 (1959) —
C.P. Conduit
IR Spectrum — Anal. Chem. <u>32</u>, 495 (1960) —
F. Pristein, M. Halik, A. Castelli
and W. Fredericks
Vibrational Analysis — Opt. Spektrosk.
<u>16</u>, 425 (1964) — A.F. Stanevich
Vibrational Analysis — J. Chim. Phys. <u>64</u>,
1450 (1967) — P. Pommez, M. Lafaix,
P. Delorme and V. Lorenzelli
Vibrational Analysis — Indian J. Phys. <u>42</u>,
511 (1968) — J.V. Shukla, V. B. Singh
and K.N. Upadhya
Raman — Indian J. Pure Appl. Phys. <u>7</u>,
830 (1969) — J.V. Shukla and K.N.
Upadhya
Vibrational Analysis — Spectrochim. Acta
<u>27A</u>, 817 (1971) — J.H.S. Green and
H.A. Lauwers
Calculation — J. Mol. Struct. <u>18</u>, 457
(1973) — A.I. Kiss and J. Szoke

2,4-Dinitrobromobenzene $(C_6H_3Br(NO_2)_2)$

Vibrational Analysis — Spectrochim. Acta
<u>20</u>, 1021 (1964) — E.F. Mooney

2,4-Dinitrochlorobenzene $(C_6H_3Cl(NO_2)_2)$

Vibrational Analysis — Spectrochim. Acta
<u>20</u>, 1021 (1964) — E.F. Mooney

1,3-Dinitronaphthalene $(C_{10}H_6(NO_2)_2)$

Phosphorescence — Spectrochim. Acta <u>27A</u>,
787 (1971) — R. Pusakowicz and A.C.
Testa

1,5-Dinitronaphthalene $(C_{10}H_6(NO_2)_2)$

Phosphorescence — Spectrochim. Acta
<u>27A</u>, 787 (1971) — R. Pusakowicz and
A.C. Testa

1,8-Dinitronaphthalene $(C_{10}H_6(NO_2)_2)$

Phosphorescence — Spectrochim. Acta 27A, 787 (1971) — R. Pusakowicz and A.C. Testa

2,4-Dinitrophenol $(C_6H_3OH(NO_2)_2)$

Raman — Indian J. Pure Appl. Phys. 11, 787 (1973) — J.V. Shukla and K.N. Upadhya

2,4-Dinitrotoluene $(C_6H_3CH_3(NO_2)_2)$

Raman — Indian J. Pure Appl. Phys. 11, 787 (1973) — J.V. Shukla and K.N. Upadhya

o-Ethoxyaniline $(C_6H_4NH_2OC_2H_5)$

IR Spectrum — Indian J. Pure Appl. Phys. 9, 376 (1971) — A.N. Pandey and N.K. Sanyal

m-Ethoxyaniline $(C_6H_4NH_2OC_2H_5)$

IR Spectrum — Indian J. Phys. 37, 139 (1963) — K.C. Medhi and G.S. Kastha

p-Ethoxyaniline $(C_6H_4NH_2OC_2H_5)$

IR Spectrum — Indian J. Pure Appl. Phys. 9, 376 (1971) — A.N. Pandey and N.K. Sanyal

1-Ethoxynaphthalene $(C_{10}H_7OC_2H_5)$

UV Absorption in Vapour — Spectrochim. Acta 30A, 277 (1974) — D. Marjit and S.B. Banerjee
IR Spectrum — Indian J. Pure Appl. Phys. 12, 727 (1974) — D. Marjit and S.B. Banerjee

N-Ethylaniline $(C_6H_5NH:C_2H_5)$

IR Spectrum — J. Amer. Chem. Soc. 47, 2192 (1925) — K.F. Bell
Raman — Monatsh. Chem. 69, 363 (1936) — L. Kahovec and A.W. Reitz
IR Spectrum — J. Chem. Soc. 843 (1957) — D. Hadzi and M. Skrbljak

UV Absorption in Vapour — Indian J. Pure Appl. Phys. 1, 193 (1963) — C.P.D. Dwivedi and D. Sharma
UV Absorption in Vapour — Indian J. Pure Appl. Phys. 6, 141 (1968) — C.P.D. Dwivedi
Vibrational Analysis — Indian J. Pure Appl. Phys. 7, 750 (1969) — C.P.D. Dwivedi
Vibrational Analysis — Indian J. Pure Appl. Phys. 9, 336 (1971) — V.N. Verma, K.P.R. Nair and D.K. Rai

o-Ethylaniline $(C_6H_4NH_2C_2H_5)$

UV Absorption in Vapour — Indian J. Pure Appl. Phys. 2, 105 (1964) — C.P.D. Dwivedi
Vibrational Analysis — Indian J. Pure Appl. Phys. 7, 750 (1969) — C.P.D. Dwivedi

Ethylbenzene $(C_6H_5C_2H_5)$

UV Absorption in Vapour — J. Chem. Phys. 13, 309 (1945) — F.A. Matsen
Vibrational Analysis — J. Chem. Phys. 13, 547 (1945) — F.G. Brickwedde, M. Moscow and R.N. Scott
UV Absorption in Solution — Chem. Rev. 41, 273 (1947) — F.A. Matsen, W.W. Robert and R.L. Chuoke
Vibrational Analysis — Izv. Akad. Nauk. SSSR 501 (1950) — P.A. Bazhulin, S.A. Ukholin, A.L. Liebermann, S.S. Nevikov and B.A. Kazanski
Vibrational Analysis — Indian J. Phys. 25, 131 (1951) — A.K. Roy
UV Absorption in Different States — Indian J. Phys. 26, 201 (1952) — A.R. Deb
Vibrational Analysis — Spectrochim. Acta 18, 39 (1962) — J.H.S. Green
Fluorescence — J. Mol. Spectrosc. 23, 365 (1967) — M.D. Lumb and D.A. Wegl
Vibrational Analysis — Spectrochim. Acta 24A, 2023 (1968) — J.E. Saunders, J.J. Lucier and J.N. Willis, Jr
Raman — J. Mol. Struct. 5, 477 (1970) — N.T. McDevitt and W.G. Fateley
IR Spectrum — J. Mol. Struct. 12, 45 (1972) — J. Stokr, H. Piveova, B. Schneider and S. Dirlikov

2-Ethyl-1,3-Dimethylbenzene
($C_6H_3C_2H_5(CH_3)_2$)

See   2-Ethyl-m-Xylene

3-Ethyl-1,2-Dimethylbenzene
($C_6H_3C_2H_5(CH_3)_2$)

See   3-Ethyl-o-Xylene

o-Ethylphenol ($C_6H_4OHC_2H_5$)

IR Spectrum - Indian J. Pure Appl. Phys.
$\underline{5}$, 242 (1967) - V.K. Mehrotra
UV Absorption in Vapour - Indian J. Pure
Appl. Phys. $\underline{6}$, 206 (1968) - V.K.
Mehrotra

m-Ethylphenol ($C_6H_4OHC_2H_5$)

IR Spectrum - Indian J. Pure Appl. Phys.
$\underline{5}$, 242 (1967) - V.K. Mehrotra
UV Absorption in Vapour - Indian J. Pure
Appl. Phys. $\underline{6}$, 206 (1968) - V.K.
Mehrotra

p-Ethylphenol ($C_6H_4OHC_2H_5$)

UV Absorption in Vapour - Curr. Sci. $\underline{35}$,
62 (1966) - M.A. Shashidhar and K.S.
Rao
IR Spectrum - Indian J. Pure Appl. Phys.
$\underline{5}$, 242 (1967) - V.K. Mehrotra
UV Absorption in Vapour - Indian J. Pure
Appl. Phys. $\underline{6}$, 206 (1968) - V.K.
Mehrotra

o-Ethylpyridine ($C_5H_4NC_2H_5$)

UV Absorption in Solution - Trans. Faraday
Soc. $\underline{50}$, 918 (1954) - R.J.L. Andon,
J.D. Cox and E.F. Herrington
IR Spectrum - Indian J. Pure Appl. Phys.
$\underline{6}$, 262 (1968) - S.M. Pandey
UV Absorption in Different States - Indian
J. Phys. $\underline{45}$, 115 (1971) - S.C. Bag

m-Ethylpyridine ($C_5H_4NC_2H_5$)

UV Absorption in Solution - Trans. Faraday
Soc. $\underline{50}$, 918 (1954) - R.J.L. Andon,
J.D. Cox and E.F. Herrington

p-Ethylpyridine ($C_5H_4NC_2H_5$)

UV Absorption in Solution - Trans. Faraday
Soc. $\underline{50}$, 918 (1954) - R.J.L. Andon,
J.D. Cox and E.F. Herrington
IR Spectrum - Indian J. Pure Appl. Phys.
$\underline{6}$, 262 (1968) - S.M. Pandey
UV Absorption in Different States - Indian
J. Phys. $\underline{45}$, 115 (1971) - S.C. Bag

o-Ethyltoluene ($C_6H_4CH_3C_2H_5$)

Vibrational Analysis - Anal. Chem. $\underline{19}$, 700
(1947) - M.R. Fenske, W.G. Braun, R.V.
Wiegand, D. Quiggle, R.M. McCormick
and D.H. Rank
Vibrational Analysis - Izv. Akad. Nauk.
SSSR $\underline{19}$, 225 (1955) - V.T. Aleksanian,
Kh.E. Sterin, A.L. Liebermann, E.A.
Mikhaileva, M.A. Prinishnikova and B.A.
Kazanski

m-Ethyltoluene ($C_6H_4CH_3C_2H_5$)

Vibrational Analysis - Angew. Chem. $\underline{A59}$,
142 (1947) - H. Fromherz and H. Bueren
Vibrational Analysis - Izv. Akad. Nauk.
SSSR $\underline{14}$, 501 (1950) - P.A. Bazhulin
S.A. Ukholin, A.L. Liebermann, S.S.
Novikov and B.A. Kazanski

p-Ethyltoluene ($C_6H_4CH_3C_2H_5$)

Vibrational Analysis - Izv. Akad. Nauk.
SSSR $\underline{14}$, 501 (1950) - P.A. Bazhulin,
S.A. Ukholin, A.L. Liebermann, S.S.
Novikov and B.A. Kazanski

2-Ethyl-m-Xylene ($C_6H_3(CH_3)_2C_2H_5$)

Vibrational Analysis - Anal. Chem. $\underline{19}$, 700
(1947) - M.R. Fenske, W.G. Braun, R.V.
Wiegand, D. Quiggle, R.M. McCormick
and D.H. Rank

4-Ethyl-m-Xylene ($C_6H_3(CH_3)_2C_2H_5$)

Vibrational Analysis - Angew. Chem. $\underline{A59}$,
142 (1947) - H. Fromherz and H. Bueren

3-Ethyl-o-Xylene $(C_6H_3(CH_3)_2C_2H_5)$

Vibrational Analysis — Anal. Chem. <u>19</u>, 700
(1947) — M.R. Fenske, W.G. Braun, R.V.
Wiegand, D. Quiggle, R.M. McCormick
and D.H. Rank

4-Ethyl-o-Xylene $(C_6H_3(CH_3)_2C_2H_5)$

Vibrational Analysis — Vestnik. Moskov
Univ. <u>17</u>, 66 (1962) — E.G. Treshova,
V.R. Skvartehenko and R.I. Levina

2-Ethyl-p-Xylene $(C_6H_3(CH_3)_2C_2H_5)$

Vibrational Analysis — Izv. Akad. Nauk.
SSSR 1444 (1961) — Kh.E. Sterin, V.T.
Aleksanian, S.A. Ukholin, O.V. Bragin,
O.D. Sterligov and B.A. Kazanski

Ethynylbenzene $(C_6H_5C:CH)$

Vibrational Analysis — J. Chem. Phys. <u>6</u>, 1
124 (1938) — J.T. Edsall and E.B. Wilson,
Jr
Vibrational Analysis — J. Chem. Soc. 597
(1944) — H.W. Thompson and P. Terkington
Vibrational Analysis — J. Opt. Soc. Amer.
<u>40</u>, 89 (1950) — D.H. Rank and R.E.
Kagarise
UV Absorption in Vapour — J. Sci. Ind. Res.
(India) <u>B19</u>, 1 (1959) — M.R. Padhye
and B.S. Rao
Vibrational Analysis — Spectrochim. Acta
<u>16</u>, 918 (1960) — J.C. Evans and R.A.
Nyquist
UV Absorption in Vapour — J. Amer. Chem.
Soc. <u>72</u>, 5260 (1960) — W.W. Robertson,
J.F. Music and F.A. Matsen
Phosphorescence — Opt. Spektrosk. <u>22</u>, 489
(1967) — T.S. Zhuravleva, R.N. Nurmu-
phamtov, Y.I. Kozlov and D.N. Shigorin
UV Absorption in Vapour — J. Mol. Spectrosc.
<u>33</u>, 376 (1970) — G.W. King and S.P. So
Vibrational Analysis — J. Mol. Spectrosc.
<u>36</u>, 468 (1970) — G.W. King and S.P. So
Calculation — J. Mol. Spectrosc. <u>37</u>, 535
(1971) — G.W. King and S.P. So
UV Absorption in Vapour — J. Mol. Spectrosc.
<u>37</u>, 543 (1971) — G.W. King and S.P. So

Ethynylbenzene-α-D_1 $(C_6H_5C:CD)$

Vibrational Analysis — Spectrochim. Acta
<u>16</u>, 918 (1960) — J.C. Evans and R.A.
Nyquist

UV Absorption in Vapour — J. Mol. Spectrosc.
<u>37</u>, 543 (1971) — G.W. King and S.P. So

Ethynylbenzene-D_5 $(C_6D_5C:CH)$

Vibrational Analysis — J. Mol. Spectrosc.
<u>36</u>, 468 (1970) — G.W. King and S.P. So

Ethynylbenzene-D_5-α-D_1 $(C_6D_5C:CD)$

Vibrational Analysis — J. Mol. Spectrosc.
<u>36</u>, 468 (1970) — G.W. King and S.P. So

o-Fluoroaniline $(C_6H_4NH_2F)$

Raman — Monatsh. Chem. <u>75</u>, 49 (1947) —
K.W.F. Kohlrausch, E. Herz and R. Vogel
Vibrational Analysis — Proc. Roy. Soc.
<u>A243</u>, 1143 (1957) — P.J. Krueger and
H.W. Thompson
IR Spectrum — Aust. J. Chem. <u>15</u>, 626
(1962) — A.N. Hamby and B.V.O. Grady
Vibrational Analysis — Can. J. Chem. <u>40</u>,
2300 (1962) — P.J. Krueger
UV Absorption in Vapour — Indian J. Pure
Appl. Phys. <u>3</u>, 495 (1965) — D.S.N.
Murty and C. Santhanam
UV Absorption in Vapour — Curr. Sci. <u>34</u>,
479 (1965) — M.A. Shashidhar and K.S.
Rao
UV Absorption in Vapour — Indian J. Phys.
<u>40</u>, 53 (1966) — M.A. Shashidhar and
K.S. Rao
IR Spectrum — Indian J. Pure Appl. Phys.
<u>4</u>, 170 (1966) — M.A. Shashidhar and
E.S. Jayadevappa
UV Emission in Vapour — Curr. Sci. <u>35</u>, 360
(1966) — S.N. Singh and N.L. Singh
IR Spectrum — Curr. Sci. <u>36</u>, 9 (1967) —
S.N. Singh and N.L. Singh
IR Spectrum — Indian J. Pure Appl. Phys.
<u>7</u>, 250 (1969) — S.N. Singh and N.L.
Singh
Vibrational Analysis — Spectrochim. Acta
<u>26A</u>, 2373 (1970) — M.A. Shashidhar,
K.S. Rao and E.S. Jayadevappa

m-Fluoroaniline $(C_6H_4NH_2F)$

Raman — Monatsh. Chem. <u>76</u>, 249 (1947) —
K.W.F. Kohlrausch
Raman — Monatsh. Chem. <u>76</u>, 49 (1947) —
K.W.F. Kohlrausch and R. Vogel
Calculation — Proc. Roy. Soc. <u>A244</u>, 143
(1957) — H.W. Thompson and P.J. Krueger

UV Absorption in Solution — Can. J. Chem. <u>36</u>, 1371 (1958) — W.E. Forbes and I.R. Leckie

Vibrational Analysis — Spectrochim. Acta <u>19</u>, 675 (1963) — G. Varsanyi, S. Holly and T. Farago

UV Absorption in Vapour — Indian J. Pure Appl. Phys. <u>3</u>, 176 (1965) — N.L. Singh and P.K. Verma

UV Absorption in Vapour — Curr. Sci. <u>34</u>, 479 (1965) — M.A. Shashidhar and K.S. Rao

UV Absorption in Vapour — Indian J. Pure Appl. Phys. <u>3</u>, 495 (1965) — D.S.N. Murty and C. Santhanam

UV Absorption in Vapour — Indian J. Phys. <u>40</u>, 53 (1966) — M.A. Shashidhar and K.S. Rao

UV Emission in Vapour — Curr. Sci. <u>35</u>, 145 (1966) — P.K. Verma

IR Spectrum — Indian J. Pure Appl. Phys. <u>4</u>, 170 (1966) — M.A. Shashidhar, K.S. Rao and E.S. Jayadevappa

IR Spectrum — Indian J. Pure Appl. Phys. <u>6</u>, 144 (1968) — P.K. Verma

Vibrational Analysis — Spectrochim. Acta <u>26A</u>, 2373 (1970) — M.A. Shashidhar, K.S. Rao and E.S. Jayadevappa

Vibrational Analysis — Indian J. Phys. <u>51B</u>, 58 (1977) — P.K. Verma

p—Fluoroaniline ($C_6H_4NH_2F$)

Raman — Monatsh. Chem. <u>66</u>, 285 (1935) — K.W. F. Kohlrausch and G.P. Ypsilanti

Raman — Monatsh. Chem. <u>76</u>, 200 (1947) — K.W.F. Kohlrausch, R. Vogel and E. Herz

UV Absorption in Vapour — Indian J. Pure Appl. Phys. <u>3</u>, 495 (1965) — D.S.N. Murty and C. Santhanam

UV Absorption in Vapour — Curr. Sci. <u>34</u>, 479 (1965) — M.A. Shashidhar and K.S. Rao

UV Absorption in Vapour — Indian J. Pure Appl. Phys. <u>3</u>, 370 (1965) — K.N. Upadhya and S.K. Tewari

UV Absorption in Vapour — Indian J. Phys. <u>40</u>, 53 (1966) — M.A. Shashidhar and K.S. Rao

IR Spectrum — Indian J. Pure Appl. Phys. <u>4</u>, 170 (1966) — M.A. Shashidhar, K.S. Rao and E.S. Jayadevappa

UV Emission in Vapour — Curr. Sci. <u>35</u>, 513 (1966) — M.A. Shashidhar and K.S. Rao

IR Spectrum — Indian J. Pure Appl. Phys. <u>6</u>, 698 (1968) — S.K. Tewari and K.N. Upadhya

UV Absorption in Vapour (High Resolution) — Mol. Phys. <u>18</u>, 451 (1970) — J. Christoffersen, J.M. Hollas and G.H. Kirby

Vibrational Analysis — Spectrochim. Acta <u>26A</u>, 2373 (1970) — M.A. Shashidhar, K.S. Rao and E.S. Jayadevappa

UV Absorption in Vapour (High Resolution) — J. Mol. Struct. <u>5</u>, 309 (1970) — S.N. Thakur, S.K. Tewari and D.K. Rai

Calculation — J. Mol. Struct. <u>14</u>, 61 (1972) — L. Smetankine and J. Etchepare

o—Fluoroanisole ($C_6H_5OCH_3F$)

Raman — Monatsh. Chem. <u>66</u>, 285 (1935) — K.W.F. Kohlrausch and G.P. Ypsilanti

Raman — Monatsh. Chem. <u>76</u>, 8 (1947) — E. Herz

UV Absorption in Vapour — Indian J. Pure Appl. Phys. <u>2</u>, 105 (1964) — L.N. Tripathi

Vibrational Analysis — Spectrochim. Acta <u>25A</u>, 343 (1969) — N.L. Owen and R.E. Hester

UV Absorption in Vapour — Indian J. Pure Appl. Phys. <u>7</u>, 357 (1969) — L.N. Tripathi

UV Emission in Vapour — J. Mol. Spectrosc. <u>34</u>, 468 (1970) — B.J. Ansari, D. Sharma and S.L. Shrivastava

Vibrational Analysis — Spectrochim. Acta <u>27A</u>, 2199 (1971) — J.H.S. Green, D.J. Harrison and W. Kynaston

m—Fluoroanisole ($C_6H_5OCH_3F$)

Vibrational Analysis — Indian J. Pure Appl. Phys. <u>7</u>, 430 (1969) — P.D. Singh

Vibrational Analysis — Spectrochim. Acta <u>25A</u>, 343 (1969) — N.L. Owen and R.E. Hester

UV Emission in Vapour — J. Mol. Spectrosc. <u>34</u>, 468 (1970) — B.J. Ansari, D. Sharma and S.L. Shrivastava

p—Fluoroanisole ($C_6H_5OCH_3F$)

IR Spectrum — J. Phys. Radium <u>9</u>, 13 (1938) — J. Lecomte

Raman — Monatsh. Chem. <u>72</u>, 244 (1939) — O. Paulsen

UV Absorption in Vapour — J. Sci. Ind. Res. (India) <u>14B</u>, 479 (1955) — V.R. Rao

UV Absorption in Vapour — Spectrochim. Acta <u>9</u>, 253 (1957) — V.R. Rao and V. Suryanarayan

Vibrational Analysis — Spectrochim. Acta
25A, 343 (1969) — N.L. Owen and R.E.
Hester
UV Emission in Vapour — Indian J. Pure Appl.
Phys. 3, 100 (1965) — K.N. Upadhya and
J.N. Rai
Fluorescence — Indian J. Pure Appl. Phys.
3, 100 (1965) — K.N. Upadhya and J.N.
Rai
Vibrational Analysis — Spectrochim. Acta
22, 1427 (1966) — J.N. Rai and K.N.
Upadhya
Vibrational Analysis — Spectrochim. Acta
23A, 1111 (1967) — M. Horak, E.R.
Lippincott and R.K. Khanna

o-Fluorobenzaldehyde (C_6H_4CHOF)

Vibrational Analysis — J. Chem. Soc. 3372
(1961) — C.J.W. Brooks and J.F. Norman
UV Absorption in Vapour — J. Sci. Ind. Res.
(India) 21B, 330 (1962) — D. Sharma
and K. Chandra
UV Absorption in Vapour — Indian J. Phys.
37, 405 (1963) — K. Chandra and D.
Sharma
UV Absorption in Vapour — Indian J. Pure
Appl. Phys. 1, 51 (1963) — M.R. Padhye
and B.G. Viladkar
Vibrational Analysis — Indian J. Pure Appl.
Phys. 4, 214 (1966) — K. Chandra
UV Emission in Vapour — Curr. Sci. 36, 399
(1967) — M.P. Srivastava and I.S. Singh
Vibrational Analysis — Curr. Sci. 36, 365
(1967) — V.B. Singh and I.S. Singh
UV Absorption in Vapour — J. Sci. Res. BHU
(India) 17, 252 (1966) — M.P. Srivas-
tava and I.S. Singh
UV Absorption in Solution — Spectrochim.
Acta 28A, 1969 (1972) — E.V. Donckt
and C. Vogels
Vibrational Analysis — Spectrochim. Acta
32A, 1265 (1976) — J.H.S. Green and
D.J. Harrison

m-Fluorobenzaldehyde (C_6H_4CHOF)

UV Absorption in Vapour — Indian J. Pure
Appl. Phys. 1, 51 (1963) — M.R. Padhye
and B.G. Viladkar
UV Absorption in Vapour — Curr. Sci. 36,
456 (1967) — M.P. Srivastava and I.S.
Singh
UV Emission in Vapour — Curr. Sci. 36, 399
(1967) — M.P. Srivastava and I.S.
Singh

Vibrational Analysis — Curr. Sci. 36, 365
(1967) — V.B. Singh and I.S. Singh
UV Absorption in Solution — Spectrochim.
Acta 28A, 1969 (1972) — E.V. Donckt
and C. Vogels
Vibrational Analysis — Spectrochim. Acta
32A, 1265 (1976) — J.H.S. Green and
D.J. Harrison

p-Fluorobenzaldehyde (C_6H_4CHOF)

UV Absorption in Vapour — Indian J. Pure
Phys. 1, 51 (1963) — M.R. Padhye and
B.G. Viladkar
UV Absorption in Vapour — Indian J. Phys.
37, 405 (1963) — K. Chandra and D.
Sharma
Vibrational Analysis — Indian J. Pure
Appl. Phys. 4, 214 (1966) — K. Chandra
Vibrational Analysis — Curr. Sci. 36, 365
(1967) — V.B. Singh and I.S. Singh
UV Emission in Vapour — Curr. Sci. 36,
399 (1967) — M.P. Srivastava and I.S.
Singh
UV Absorption in Vapour — J. Sci. Res. BHU
(India) 17, 252 (1966) — M.P. Srivas-
stava and I.S. Singh
Raman — Spectrochim. Acta 23A, 462 (1967) —
F.B. Brown
UV Absorption in Solution — Spectrochim.
Acta 28A, 1969 (1972) — E.V. Donckt
and C. Vogels
Vibrational Analysis — Spectrochim. Acta
32A, 1265 (1976) — J.H.S. Green and
D.J. Harrison

Fluorobenzene (C_6H_5F)

Raman — Monatsh. Chem. 74, 1 (1941) —
K.W.F. Kohlrausch and H. Wittek
Vibrational Analysis — Proc. Indian Acad.
Sci. A15, 401 (1942) — C.S. Venkate-
shwaran and N.S. Pandya
UV Absorption in Vapour — J. Chem. Phys.
14, 123 (1946) — S.H. Wellman
Fluorescence — J. Chem. Phys. 18, 1403
(1950) — A.M. Bass
UV Absorption in Vapour — J. Opt. Soc.
Amer. 40, 389 (1950) — A.M. Bass
UV Absorption in Vapour — Phys. Rev. 87,
213 (1952) — V.R. Rao and H. Sponer
Vibrational Analysis — J. Chem. Phys. 21,
1457 (1953) — D.C. Smith, E.E.
Ferguson, R.L. Hudson and J.R. Nielsen
Vibrational Analysis — J. Chem. Phys. 21,
1727 (1953) — D.C. Smith, E.E. Ferguson,
R.L. Hudson and J.R. Nielsen

70

Vibrational Analysis — J. Chem. Phys. <u>21</u>, 1475 (1953) — C.H. Smith

IR Spectrum — J. Chem. Soc. 1350 (1956) — D.H. Whiffen

Vibrational Analysis — J. Amer. Chem. Soc. <u>78</u>, 5457 (1956) — D.W. Scott

UV Absorption in Solution — Can. J. Chem. <u>37</u>, 1977 (1959) — W.F. Forbes

IR Spectrum — J. Sci. Ind. Res. (India) <u>18B</u>, 504 (1959) — M.R. Padhye and B.G. Viladkar

Raman — Indian J. Phys. <u>34</u>, 402 (1960) — K.K. Deb

IR Spectrum — J. Chem. Soc. 2236 (1961) — J.H.S. Green

IR Spectrum — Indian J. Phys. <u>38</u>, 610 (1964) — S.C. Sirkar, D.K. Mukherjee and P.K. Bishnui

Fluorescence — J. Phys. Chem. <u>69</u>, 4284 (1965) — I. Unger

UV Absorption in Different States — J. Chem. Phys. <u>62</u>, 1242 (1965) — C. Bacharan and J. Kahane-Pailloue

IR Spectrum — Spectrochim. Acta <u>22</u>, 501 (1966) — W.R. McWhinnie and R.C. Poller

Fluorescence — J. Phys. Chem. <u>71</u>, 1839 (1967) — D. Phillips

Fluorescence — J. Chem. Phys. <u>42</u>, 2942 (1968) — G.B. Kistiakowsky and C.S. Parmenter

UV Absorption in Vapour — Mol. Phys. <u>19</u>, 289 (1970) — G.H. Kirby

Fluorescence — J. Chem. Phys. <u>53</u>, 998 (1970) — K. Nakamura

Raman — J. Mol. Struct. <u>5</u>, 477 (1970) — N.T. McDevitt and W.G. Fateley

Fluorescence — J. Phys. Chem. <u>75</u>, 3662 (1971) — K. Al-Ani and D. Phillips

Fluorescence — J. Chem. Phys. <u>56</u>, 2291 (1972) — A.S. Abramson, K.G. Spears and S.A. Rice

Fluorescence — Chem. Phys. Lett. <u>13</u>, 140 (1972) — G.L. Loper and E.K.C. Lee

Fluorescence — Chem. Phys. Lett. <u>14</u>, 404 (1972) — G.M. Breuer and E.K.C. Lee

Calculation — J. Mol. Spectrosc. <u>48</u>, 446 (1973) — V.J. Eaton and D. Steele

Calculation — J. Mol. Struct. <u>16</u>, 365 (1973) — R.T.C. Brownlee, D.G. Cameron, R.D. Topsom, A.R. Katritzky and A.J. Sparrow

Calculation — J. Mol. Struct. <u>32</u>, 93 (1976) — F. Torok, A. Hegedus, K. Kosa and P. Pulay

Fluorobenzene-D$_5$ (C$_6$D$_5$F)

Vibrational Analysis — J. Chem. Phys. <u>33</u>, 1242 (1960) — D. Steele, E.R. Lippincott and J. Xavier

Calculation — J. Mol. Spectrosc. <u>48</u>, 446 (1973) — V.J. Eaton and D. Steele

o-Fluorobenzoic acid (C$_6$H$_4$HCO$_2$F)

Vibrational Analysis — Spectrochim. Acta <u>33A</u>, 575 (1977) — J.H.S. Green

m-Fluorobenzoic acid (C$_6$H$_4$HCO$_2$F)

Vibrational Analysis — Spectrochim. Acta <u>33A</u>, 575 (1977) — J.H.S. Green

p-Fluorobezoic acid (C$_6$H$_4$HCO$_2$F)

Vibrational Analysis — Spectrochim. Acta <u>33A</u>, 575 (1977) — J.H.S. Green

o-Fluorobenzonitrile (C$_6$H$_4$CNF)

Vibrational Analysis — Indian J. Pure Appl. Phys. <u>7</u>, 430 (1969) — P.D. Singh

Fluorescence — J. Lumin. <u>9</u>, 449 (1975) — Y.H. Lui and S.P. McGlynn

Vibrational Analysis — Spectrochim. Acta <u>32A</u>, 1279 (1976) — J.H.S. Green and D.J. Harrison

IR Spectrum — Indian J. Phys. <u>50</u>, 478 (1976) — B.B. Lal and I.S. Singh

m-Fluorobenzonitrile (C$_6$H$_4$CNF)

Vibrational Analysis — Spectrochim. Acta <u>19</u>, 807 (1963) — J.H.S. Green, W. Kynaston and H.A. Gebbie

UV Absorption in Vapour — Indian J. Phys. <u>42</u>, 571 (1968) — P.D. Singh

Vibrational Analysis — Indian J. Pure Appl. Phys. <u>7</u>, 430 (1969) — P.D. Singh

Fluorescence — J. Lumin. <u>9</u>, 449 (1975) — Y.H. Lui and S.P. McGlynn

Vibrational Analysis — Spectrochim. Acta <u>32A</u>, 1279 (1976) — J.H.S. Green and D.J. Harrison

p—Fluorobenzonitrile (C_6H_4CNF)

UV Absorption in Vapour — J. Chem. Phys.
21, 379 (1953) — C.D. Cooper
IR Spectrum — Indian J. Pure Appl. Phys. 4,
169 (1966) — S.M. Pandey and B.R. Pandey
UV Absorption in Vapour — Indian J. Pure
Appl. Phys. 4, 169 (1966) — S.M. Pandey
and B.R. Pandey
UV Absorption in Vapour — Indian J. Pure
Appl. Phys. 9, 109 (1971) — B.R. Pandey
and D. Sharma
IR Spectrum — Indian J. Pure Appl. Phys.
9, 109 (1971) — B.R. Pandey and D.
Sharma

p—Fluorobenzoylchloride (C_6H_4FCOCl)

IR Spectrum — Indian J. Pure Appl. Phys.
12, 461 (1974) — G.N.R. Tripathi, D.R.
Singh and U.S. Tripathi
UV Absorption in Vapour — Indian J. Pure
Appl. Phys. 12, 461 (1974) — G.N.R.
Tripathi, D.R. Singh and U.S. Tripathi

o—Fluorobenzotrifluoride ($C_6H_4FCF_3$)

Fluorescence — J. Chem. Phys. 55, 5753
(1971) — D. Gray and D. Phillips
Fluorescence — J. Chem. Phys. 58, 5073
(1973) — K. Al-Ani
Fluorescence — J. Chem. Phys. 59, 330
(1973) — K. Al-Ani
Fluorescence — J. Chem. Phys. 59, 341
(1973) — K. Al-Ani
UV Absorption in Vapour — Indian J. Pure
Appl. Phys. 11, 705 (1973) — P.D. Singh
and G.C. Singh
Vibrational Analysis — Spectrochim. Acta
33A, 837 (1977) — J.H.S. Green and
D.J. Harrison

m—Fluorobenzotrifluoride ($C_6H_4FCF_3$)

UV Absorption in Vapour — J. Chem. Phys.
17, 845 (1949) — H.W. Thompson and
C.H. Miller
Fluorescence — J. Chem. Phys. 55, 5753
(1971) — D. Gray and D. Phillips
Fluorescence — J. Chem. Phys. 58, 5073
(1973) — K. Al-Ani
Fluorescence — J. Chem. Phys. 59, 330
(1973) — K. Al-Ani
Fluorescence — J. Chem. Phys. 59, 341
(1973) — K. Al-Ani

UV Absorption in Vapour — Indian J. Pure
Appl. Phys. 11, 705 (1973) — P.D.
Singh and G.C. Singh
Vibrational Analysis — Spectrochim. Acta
33A, 837 (1977) — J.H.S. Green and
D.J. Harrison

p—Fluorobenzotrifluoride ($C_6H_4FCF_3$)

Vibrational Analysis — J. Chem. Soc. 1432
(1948) — H.W. Thompson and N.N. Temple
UV Absorption in Vapour — J. Chem. Phys.
17, 845 (1949) — H.W. Thompson and
C.H. Miller
Fluorescence — J. Chem. Phys. 55, 5753
(1971) — D. Gray and D. Phillips
Fluorescence — J. Chem. Phys. 58, 5073
(1973) — K. Al-Ani
Fluorescence — J. Chem. Phys. 59, 341
(1973) — K. Al-Ani
Vibrational Analysis — Spectrochim. Acta
33A, 837 (1977) — J.H.S. Green and
D.J. Harrison

p—Fluorobenzylbromide ($C_6H_4FCH_2Br$)

Vibrational Analysis — Spectrochim. Acta
28A, 55 (1972) — L. Verdonck and G.P.
Van der Kelen

m—Fluorobenzylchloride ($C_6H_4FCH_2Cl$)

Vibrational Analysis — Spectrochim. Acta
29A, 813 (1973) — L. Verdonck, G.P.
Van der Kelen and Z. Eeckhaut

p—Fluorobenzylchloride ($C_6H_4FCH_2Cl$)

Vibrational Analysis — Spectrochim. Acta
28A, 55 (1972) — L. Verdonck and G.P.
Van der Kelen

2—Fluoro—4—Bromotoluene ($C_6H_3BrFCH_3$)

IR Spectrum — Indian J. Pure Appl. Phys.
11, 446 (1973) — C.P.D. Dwivedi and
S.N. Sharma
UV Absorption in Vapour — Curr. Sci. 42,
784 (1973) — C.P.D. Dwivedi and S.N.
Sharma

2-Fluoro-5-Bromotoluene $(C_6H_3BrFCH_3)$

IR Spectrum — Indian J. Pure Appl. Phys.
  11, 446 (1973)   C.P.D. Dwivedi and
  S.N. Sharma
UV Absorption in Vapour — Curr. Sci. **42**,
  784 (1973) — C.P.D. Dwivedi and S.N.
  Sharma
Vibrational Analysis — Indian J. Pure Appl.
  Phys. **16**, 532 (1978) — S.N. Singh,
  R.B. Gupta, P.L. Gupta and C.P.D.
  Dwivedi

3-Fluoro-6-Bromotoluene $(C_6H_3BrFCH_3)$

UV Absorption in Vapour — Indian J. Pure
  Appl. Phys. **10**, 492 (1972) — I.D.
  Singh and R.C. Maheshwari

4-Fluoro-2-Bromotoluene $(C_6H_3BrFCH_3)$

IR Spectrum — Indian J. Pure Appl. Phys.
  11, 446 (1973) — C.P.D. Dwivedi and
  S.N. Sharma
UV Absorption in Vapour — Curr. Sci. **42**,
  784 (1973) — C.P.D. Dwivedi and S.N.
  Sharma

4-Fluoro-3-Bromotoluene $(C_6H_3BrFCH_3)$

UV Absorption in Vapour — Indian J. Pure
  Appl. Phys. **10**, 492 (1972) — I.D.
  Singh and R.C. Maheshwari

3-Fluoro-4-Chloroaniline $(C_6H_3NH_2FCl)$

Vibrational Analysis — Indian J. Pure Appl.
  Phys. **16**, 719 (1978) — N.K. Sanyal,
  S.L. Srivastava and R.K. Goel

4-Fluoro-3-Chloroaniline $(C_6H_3NH_2FCl)$

UV Absorption in Vapour — Indian J. Pure
  Appl. Phys. **6**, 229 (1968) — C.G. Rama
  Rao and C. Santhamma

2-Fluoro-1,3-Dichlorobenzene $(C_6H_3FCl_2)$

Vibrational Analysis — Spectrochim. Acta
  27A, 793 (1971) — J.H.S. Green, D.J.
  Harrison and W. Kynaston

1-Fluoro-2,3-Dimethylbenzene $(C_6H_3F(CH_3)_2)$

Vibrational Analysis — Spectrochim. Acta
  27A, 793 (1971) — J.H.S. Green, D.J.
  Harrison and W. Kynaston

1-Fluoro-2,4-Dimethylbenzene $(C_6H_3F(CH_3)_2)$

Vibrational Analysis — Indian J. Phys.
  35, 583 (1961) — K.C. Medhi
Vibrational Analysis — Spectrochim. Acta
  35A, 35 (1979) — A.K. Ansari and
  P.K. Verma

1-Fluoro-2,6-Dimethylbenzene $(C_6H_3F(CH_3)_2)$

Vibrational Analysis — Spectrochim. Acta
  27A, 793 (1971) — J.H.S. Green, D.J.
  Harrison and W. Kynaston

1-Fluoro-3,4-Dimethylbenzene $(C_6H_3F(CH_3)_2)$

Vibrational Analysis — Spectrochim. Acta
  27A, 807 (1971) — J.H.S. Green, D.J.
  Harrison and W. Kynaston

1-Fluoro-3,5-Dimethylbenzene $(C_6H_3F(CH_3)_2)$

Vibrational Analysis — Spectrochim. Acta
  27A, 793 (1971) — J.H.S. Green, D.J.
  Harrison and W. Kynaston

1-Fluoro-2,4-Dinitrobenzene $(C_6H_3F(NO_2)_2)$

Vibrational Analysis — Indian J. Phys.
  35, 583 (1961) — K.C. Medhi

o-Fluoroiodobenzene (C_6H_4FI)

UV Absorption in Vapour — Indian J. Phys.
  31, 387 (1957) — S.L.N.G. Krishnamachari
Vibrational Analysis — Curr. Sci. **26**, 144
  (1957) — S.L.N.G. Krishnamachari
Vibrational Analysis — Spectrochim. Acta
  26A, 1913 (1970) — J.H.S. Green

m-Fluoroiodobenzene (C_6H_4FI)

Vibrational Analysis — Curr. Sci. 25, 260
(1956) — S.L.N.G. Krishnamachari
Vibrational Analysis — Curr. Sci. 26, 144
(1957) — S.L.N.G. Krishnamachari
UV Absorption in Vapour — Indian J. Phys.
31, 387 (1957) — S.L.N.G. Krishnamachari
Vibrational Analysis — Spectrochim. Acta
26A, 1523 (1970) — J.H.S. Green

p-Fluoroiodobenzene (C_6H_4FI)

UV Absorption in Solution — Z. Phys. Chem.
33, 311 (1936) — H. Conrad-Billroth
and G. Forster
Vibrational Analysis — J. Chem. Phys. 24,
420 (1956) — N.A. Narasimhan, M.Z.
El-Sabban and J.R. Nielsen
UV Absorption in Vapour — Indian J. Phys.
31, 387 (1957) — S.L.N.G. Krishnamachari
Vibrational Analysis — Spectrochim. Acta
12, 57 (1958) — A. Stojilkovic and D.H.
Whiffen
Vibrational Analysis — Proc. Roy. Soc.
A298, 51 (1967) — P.R. Griffiths and
H.W. Thompson
Vibrational Analysis — J. Chim. Phys. 64,
1450 (1967) — P. Pommez, M. Lafaix and
P. Delorme
Vibrational Analysis — J. Mol. Spectrosc.
23, 44 (1968) — N.A. Narasimhan and
C.V.S.R. Rao
Vibrational Analysis — Spectrochim. Acta
26A, 1503 (1970) — J.H.S. Green

2-Fluoro-N-Methylaniline ($C_6H_4FNH:CH_3$)

IR Spectrum — Indian J. Pure Appl. Phys.
5, 141 (1967) — U. Kumar

2-Fluoro-5-Methylaniline ($C_6H_3FCH_3NH_2$)

IR Spectrum — Indian J. Pure Appl. Phys.
13, 570 (1975) — S.N. Sharma and C.P.D.
Dwivedi
UV Absorption in Vapour — Indian J. Pure
Appl. Phys. 14, 839 (1976) — S.N. Sharma,
R.P. Sinha and C.P.D. Dwivedi

3-Fluoro-N-Methylaniline ($C_6H_4FNH:CH_3$)

IR Spectrum — Indian J. Pure Appl. Phys.
5, 141 (1967) — U. Kumar

3-Fluoro-4-Methylaniline ($C_6H_3FCH_3NH_2$)

IR Spectrum — Indian J. Pure Appl. Phys.
13, 570 (1975) — S.N. Sharma and C.P.D.
Dwivedi
UV Absorption in Vapour — Indian J. Pure
Appl. Phys. 14, 839 (1976) — S.N.
Sharma, R.P. Sinha and C.P.D. Dwivedi

4-Fluoro-N-Methylaniline ($C_6H_4FNH:CH_3$)

IR Spectrum — Indian J. Pure Appl. Phys.
5, 141 (1967) — U. Kumar

α-Fluoronaphthalene ($C_{10}H_7F$)

Raman — Z. Elektrochem. 52, 210 (1948) —
V.H. Luther
IR Spectrum — J. Chem. Soc. 3645 (1954) —
J. Ferguson and R.L. Werner
IR Spectrum — J. Chem. Soc. 3160 (1954) —
J. Ferguson, T. Iredale and J.A. Tayler
UV Absorption in Solution — J. Chem. Soc.
304 (1954) — J. Ferguson
UV Absorption in Vapour — J. Sci. Ind. Res.
(India) 15B, 262 (1956) — S. Ramamurty
and V.R. Rao
UV Absorption in Vapour — Indian J. Phys.
31, 497 (1957) — S. Ramamurty, M.J.
Rao and V.R. Rao
UV Absorption in Different States — Indian
J. Phys. 34, 61 (1959) — S.B. Banerjee
IR Spectrum — Indian J. Phys. 36, 557
(1962) — K.K. Deb
UV Emission in Vapour — Indian J. Pure
Appl. Phys. 8, 179 (1970) — B. Singh,
R.D. Singh and R.M.P. Jaiswal
Vibrational Analysis — Indian J. Pure Appl.
Phys. 9, 31 (1971) — B. Singh, R.M.P.
Jaiswal and R.D. Singh
UV Absorption in Vapour — Indian J. Pure
Appl. Phys. 14, 583 (1976) — R.D. Singh
and S.N. Singh
UV Emission in Vapour — Indian J. Pure
Appl. Phys. 14, 583 (1976) — R.D. Singh
and S.N. Singh
Vibrational Analysis — Spectrochim. Acta
34A, 985 (1978) — S.N. Singh, H.S.
Bhatti and R.D. Singh

β-Fluoronaphthalene ($C_{10}H_7F$)

UV Absorption in Solution — J. Chem. Soc.
304 (1954) — J. Ferguson
IR Spectrum — J. Chem. Soc. 3645 (1954) —
J. Ferguson and R.L. Werner

IR Spectrum — J. Chem. Soc. 3160 (1954) — J. Ferguson, T. Iredale and J.A. Taylor

UV Absorption in Vapour — J. Sci. Ind. Res. (India) 14B, 547 (1955) — M.J. Rao and V.R. Rao

IR Spectrum — J. Chem. Phys. 25, 229 (1956) — J.W. Sidman

UV Absorption in Vapour — Indian J. Phys. 31, 497 (1957) — S. Ramamurty, M.J. Rao and V.R. Rao

UV Absorption in Vapour — Indian J. Pure Appl. Phys. 6, 398 (1968) — R.D. Singh

UV Emission in Vapour — Indian J. Pure Appl. Phys. 6, 398 (1968) — R.D. Singh

Vibrational Analysis — Indian J. Pure Appl. Phys. 9, 31 (1971) — B. Singh, R.M.P. Jaiswal and R.D. Singh

UV Absorption in Vapour — Indian J. Pure Appl. Phys. 13, 48 (1975) — R.D. Singh

UV Emission in Vapour — Indian J. Pure Appl. Phys. 13, 48 (1975) — R.D. Singh

o-Fluoronitrobenzene $(C_6H_4FNO_2)$

Vibrational Analysis — Spectrochim. Acta 20, 675 (1964) — K.C. Medhi

Vibrational Analysis — Spectrochim. Acta 26A, 1923 (1970) — J.H.S. Green and D.J. Harrison

m-Fluoronitrobenzene $(C_6H_4FNO_2)$

Vibrational Analysis — Spectrochim. Acta 20, 675 (1964) — K.C. Medhi

Vibrational Analysis — Spectrochim. Acta 26A, 1925 (1970) — J.H.S. Green and D.J. Harrison

p-Fluoronitrobenzene $(C_6H_4FNO_2)$

Raman — Z. Phys. Chem. 52B, 315 (1942) — H. Wittek

UV Absorption in Solution — Can. J. Chem. 31, 1020 (1953) — W. Gruber

UV Absorption in Solution — J. Amer. Chem. Soc. 76, 160 (1954) — H.E. Ungnade

UV Absorption in Solution — J. Org. Chem. 22, 1285 (1957) — W.M. Schubert, J.M. Craven, H. Steadly and J. Robins

UV Absorption in Solution — J. Amer. Chem. Soc. 81, 269 (1958) — W.M. Schubert, J.M. Craven and H. Steadly

UV Absorption in Vapour — Indian J. Phys. 36, 213 (1962) — I.A. Rao

Vibrational Analysis — Spectrochim. Acta 20, 675 (1964) — K.C. Medhi

Vibrational Analysis — Proc. Roy. Soc. A298, 51 (1967) — P.R. Griffiths and H.W. Thompson

Vibrational Analysis — J. Chim. Phys. 64, 1450 (1967) — P. Pommez, M. Lafaix, P. Delorme and V. Lorenzelli

Vibrational Analysis — Spectrochim. Acta 26A, 1925 (1970) — J.H.S. Green and D.J. Harrison

Calculation — J. Mol. Struct. 14, 61 (1972) — L. Smelankine and J. Etchepare

o-Fluorophenetole $(C_6H_4OC_2H_5F)$

IR Spectrum — Indian J. Pure Appl. Phys. 13, 135 (1975) — R.C. Maheshwari and M.M. Shukla

UV Absorption in Vapour — Indian J. Pure Appl. Phys. 13, 135 (1975) — R.C. Maheshwari and M.M. Shukla

o-Fluorophenol (C_6H_4OHF)

UV Absorption in Solution — J. Chem. Soc. 380 (1943) — H.H. Hodgen

Raman — Monatsh. Chem. 76, 249 (1947) — K.W.F. Kohlrausch and E. Herz

UV Absorption in Solution — Can. J. Chem. 37, 1294 (1959) — J.R. Dearden and W.F. Forbes

UV Absorption in Vapour — Nature 200, 1202 (1963) — S.K. Tewari

UV Emission in Vapour — Indian J. Phys. 44, 128 (1970) — K.N. Upadhya and S.K. Tewari

Vibrational Analysis — Spectrochim. Acta 27A, 2199 (1971) — J.H.S. Green, D.J. Harrison and W. Kynaston

UV Absorption in Vapour — J. Mol. Spectrosc. 37, 486 (1971) — G.N.R. Tripathi

o-Fluorophenol-OD (C_6H_4ODF)

Vibrational Analysis — Spectrochim. Acta 27A, 2199 (1971) — J.H.S. Green, D.J. Harrison and W. Kynaston

m-Fluorophenol (C_6H_4OHF)

UV Absorption in Solution — J. Chem. Soc. 380 (1943) — H.H. Hodgen

Raman — Monatsh. Chem. 76, 249 (1947) — E. Herz and K.W.F. Kohlrausch

UV Absorption in Solution — Can. J. Chem.
37, 1294 (1959) — J.R. Dearden and
W.F. Forbes
UV Absorption in Vapour — Nature 200, 1202
(1963) — S.K. Tewari
UV Emission in Vapour — Curr. Sci. 37, 160
(1968) — B.J. Ansari and S.L. Srivastava
IR Spectrum — Indian J. Pure Appl. Phys.
7, 517 (1969) — G.N.R. Tripathi
UV Absorption in Vapour — Indian J. Pure
Appl. Phys. 8, 157 (1970) — G.N.R.
Tripathi
UV Emission in Vapour — Indian J. Phys. 44,
128 (1970) — J.V. Shukla, K.N. Upadhya
and S.K. Tewari
Vibrational Analysis — Spectrochim. Acta
27A, 2199 (1971) — J.H.S. Green, D.J.
Harrison and W. Kynaston

p-Fluorophenol (C_6H_4OHF)

Raman — Monatsh. Chem. 66, 285 (1935) —
K.W.F. Kohlrausch and G.P. Ypsilanti
Raman — Sber. Akad. Wiss. Wien 144, 417
(1935) — K.W.F. Kohlrausch and G.P.
Ypsilanti
UV Absorption in Solution — J. Chem. Soc.
380 (1943) — H.H. Hodgen
UV Absorption in Solution — Can. J. Chem.
37, 1294 (1959) — J.R. Dearden and
W.F. Forbes
Vibrational Analysis — Appl. Spectrosc.
16, 32 (1962) — R.J. Jackobsen and E.J.
Brewer
UV Absorption in Vapour — Nature 200, 1202
(1963) — S.K. Tewari
UV Absorption in Vapour — Indian J. Pure
Appl. Phys. 1, 229 (1963) — D. Sharma
and L.N. Tripathi
UV Emission in Vapour — Indian J. Pure
Appl. Phys. 6, 151 (1968) — S.L.
Srivastava and L.N. Tripathi
UV Absorption in Vapour — J. Chem. Phys.
48, 348 (1968) — K.T. Huang and J.R.
Lombardi
UV Absorption in Vapour (High Resolution) —
Indian J. Pure Appl. Phys. 7, 570
(1969) — S.N. Thakur and S.K. Tewari
UV Emission in Vapour — Indian J. Phys.
44, 128 (1970) — J.V. Shukla, K.N.
Upadhya and S.K. Tewari
UV Absorption in Vapour (High Resolution) —
Mol. Phys. 18, 451 (1970) — J. Christo-
ffersen, J.M. Hollas and G.H. Kirby
Vibrational Analysis — Spectrochim. Acta
27A, 2199 (1971) — J.H.S. Green, D.J.
Harrison and W. Kynaston
IR Spectrum — J. Mol. Struct. 22, 29 (1974)—
N.W. Larsen and F.M. Nicolaisen

p-Fluorophenol-OD (C_6H_4ODF)

IR Spectrum — J. Mol. Struct. 22, 29
(1974) — N.W. Larsen and F.M. Nicolaisen

o-Fluoropyridine (C_5H_4NF)

Vibrational Analysis — Spectrochim. Acta
19, 549 (1963) — J.H.S. Green, W.
Kynaston and H.M. Paisley
Vibrational Analysis — Appl. Spectrosc.
17, 90 (1963) — R. Issac, F.F. Bentley,
H. Sternglanz, W.C. Coburn, C.V.
Stephenson and W.S. Wilcox
Vibrational Analysis — Spectrochim. Acta
33A, 75 (1977) — J.H.S. Green and D.J.
Harrison

m-Fluoropyridine (C_5H_4NF)

Vibrational Analysis — Spectrochim. Acta
19, 549 (1963) — J.H.S. Green, W.
Kynaston and H.M. Paisley

p-Fluoropyridine (C_5H_4NF)

Vibrational Analysis — Spectrochim. Acta
19, 549 (1963) — J.H.S. Green, W.
Kynaston and H.M. Paisley

o-Fluoropyrimidine ($C_4H_3N_2F$)

Vibrational Analysis — Spectrochim. Acta
33A, 189 (1977) — E. Allenstein, P.
Kiemle, J. Weidlein and W. Podszun

o-Fluorostyrene ($C_6H_4CH:CH_2F$)

UV Absorption in Vapour — Indian J. Pure
Appl. Phys. 5, 187 (1967) — B.J.
Ansari, D. Sharma and G.N.R. Tripathi
Vibrational Analysis — Appl. Spectrosc.
22, 650 (1968) — W.G. Fateley, G.L.
Carlson and F.E. Dickson
IR Spectrum — Indian J. Pure Appl. Phys.
7, 757 (1969) — G.N.R. Tripathi
Raman — Indian J. Pure Appl. Phys. 9, 199
(1971) — B. Singh and R.M.P. Jaiswal
UV Emission in Vapour — Indian J. Pure
Appl. Phys. 9, 491 (1971) — B. Singh
and R.M.P. Jaiswal

m-Fluorostyrene $(C_6H_4CH:CH_2F)$

UV Absorption in Vapour - Indian J. Pure Appl. Phys. 5, 187 (1967) - B.J. Ansari, D. Sharma and G.N.R. Tripathi

Vibrational Analysis - Appl. Spectrosc. 22, 650 (1968) - W.G. Fateley, G.L. Carlson and F.E. Dickson

IR Spectrum - Indian J. Pure Appl. Phys. 7, 757 (1969) - G.N.R. Tripathi

Raman - Indian J. Pure Appl. Phys. 9, 199 (1971) - B. Singh and R.M.P. Jaiswal

UV Emission in Vapour - Indian J. Pure Appl. Phys. 9, 491 (1971) - B. Singh and R.M.P. Jaiswal

Raman - J. Mol. Struct. 37, 85 (1977) - L.A. Carreira and T.G. Towns

p-Fluorostyrene $(C_6H_4CH:CH_2F)$

Vibrational Analysis - Vestnik. Moskov Univ. 17, 66 (1962) - E.G. Treshova, V.R. Skvartechenke and R.I. Levine

UV Absorption in Vapour - Indian J. Pure Appl. Phys. 5, 187 (1967) - B.J. Ansari, D. Sharma and G.N.R. Tripathi

IR Spectrum - Indian J. Pure Appl. Phys. 6, 519 (1968) - M.P. Srivastava, O.N. Singh and I.S. Singh

IR Spectrum - Indian J. Pure Appl. Phys. 7, 757 (1969) - G.N.R. Tripathi

Raman - Indian J. Pure Appl. Phys. 9, 199 (1971) - B. Singh and R.M.P. Jaiswal

UV Emission in Vapour - Indian J. Pure Appl. Phys. 9, 491 (1971) - B. Singh and R.M.P. Jaiswal

Raman - J. Mol. Struct. 37, 85 (1977) - L.A. Carreira and T.G. Towns

p-Fluorothiophenol (C_6H_4SHF)

Vibrational Analysis - Spectrochim. Acta 26A, 1515 (1970) - J.H.S. Green, D.J. Harrison, W. Kynaston and D.W. Scott

o-Fluorotoluene $(C_6H_4CH_3F)$

Raman - Monatsh. Chem. 76, 12 (1947) - E. Herz

Raman - J. Chem. Soc. 1432 (1948) - H.W. Thompson and N.N. Temple

UV Absorption in Solution - Disc. Faraday Soc. 9, 35 (1950) - W.T. Cave and H.W. Thompson

UV Absorption in Different States - Indian J. Phys. 34, 237 (1960) - S.K. Sen

UV Absorption in Vapour - Indian J. Phys. 35, 628 (1961) - J.K. Roy

Vibrational Analysis - Indian J. Phys. 36, 59 (1962) - K.K. Deb

Raman - J. Chim. Phys. 62, 347 (1965) - M. Brigodiot and J.M. Lebas

UV Absorption in Vapour - Indian J. Pure Appl. Phys. 4, 40 (1966) - G. Joshi

UV Emission in Vapour - Curr. Sci. 35, 512 (1966) - G. Joshi

IR Spectrum - Spectrochim. Acta 23A, 1341 (1967) - G. Joshi and N.L. Singh

Vibrational Analysis - Spectrochim. Acta 26A, 1913 (1970) - J.H.S. Green

Fluorescence - J. Phys. Chem. 75, 3214 (1971) - K. Al-Ani and D. Phillips

Calculation - J. Mol. Struct. 8, 319 (1971) - P.C. Mishra, S.N. Thakur and D.K. Rai

Fluorescence - Chem. Phys. Lett. 14, 404 (1972) - G.M. Brewer and E.K.C. Lee

m-Fluorotoluene $(C_6H_4CH_3F)$

Raman - Monatsh. Chem. 74, 166 (1941) - E. Herz

Vibrational Analysis - J. Chem. Soc. 1412 (1948) - H.W. Thompson and N.N. Temple

UV Absorption in Solution - Disc. Faraday Soc. 9, 35 (1950) - W.T. Cave and H.W. Thompson

UV Absorption in Vapour - Trans. Faraday Soc. 53, 1570 (1957) - V. Suryanarayanan

UV Absorption in Different States - Indian J. Phys. 34, 237 (1960) - S.K. Sen

UV Absorption in Solution - Indian J. Phys. 34, 331 (1960) - J.K. Roy

UV Absorption in Vapour - Indian J. Phys. 35, 628 (1961) - J.K. Roy

UV Absorption in Solution - Indian J. Phys. 36, 156 (1962) - J.K. Roy

Vibrational Analysis - Indian J. Phys. 36, 59 (1962) - K.K. Deb

UV Absorption in Vapour - J. Sci. Res. BHU (India) 15, 90 (1964) - G. Joshi and R.N. Singh

Vibrational Analysis - J. Chim. Phys. 63, 552 (1966) - C. Garrigou-Lagrange, M. Chehata and J. Lascombe

Vibrational Analysis - Spectrochim. Acta 26A, 1523 (1970) - J.H.S. Green

Fluorescence - J. Phys. Chem. 75, 3214 (1971) - K. Al-Ani and D. Phillips

Calculation - J. Mol. Struct. 8, 319 (1971) - P.C. Mishra, S.N. Thakur and D.K. Rai

Fluorescence — Chem. Phys. Lett. 14, 404
    (1972) — G.M. Brewer and E.K.C. Lee
Vibrational Analysis — Spectrochim. Acta
    29A, 813 (1973) — L. Verdonck, G.P.
    Van der Kelen and Z. Eeckhaut

p-Fluorotoluene (C_6H_4CH_3F)

$$p\text{-Fluorotoluene} \quad (C_6H_4CH_3F)$$

Vibrational Analysis — J. Chem. Soc. 1432
    (1948) — H.W. Thompson and N.N. Temple
UV Absorption in Solution — J. Chem. Phys.
    17, 845 (1949) — C.H. Miller and H.W.
    Thompson
UV Absorption in Solution — Disc. Faraday
    Soc. 9, 41 (1950) — W.T. Cave and
    H.W. Thompson
Raman — J. Chem. Phys. 21, 1736 (1953) —
    E.E. Ferguson, R.L. Hudson and J.R.
    Nielsen
Vibrational Analysis — Spectrochim. Acta
    12, 305 (1958) — C. Garrigou-Lagrange,
    J.M. Lebas and M.L. Josien
Vibrational Analysis — Spectrochim. Acta
    15, 225 (1959) — J.M. Lebas, C.
    Garrigou-Lagrange and M.L. Josien
UV Absorption in Different States — Indian
    J. Phys. 34, 237 (1960) — S.K. Sen
UV Absorption in Vapour — Indian J. Phys.
    35, 628 (1961) — J.K. Roy
Raman — Indian J. Phys. 35, 16 (1961) —
    K.K. Deb
Vibrational Analysis — Indian J. Phys.
    36, 59 (1962) — K.K. Deb
Vibrational Analysis — J. Chem. Phys. 37,
    867 (1962) — D.W. Scott, J.F. Messer-
    ley, S.S. Todd, I.A. Hossenlopp, D.R.
    Doulson and J.P. McCullough
IR Spectrum — Indian J. Phys. 39, 537
    (1965) — D.K. Mukherjee, P.K. Bishnui
    and S.C. Sirkar
Vibrational Analysis — J. Chim. Phys. 64,
    1450 (1967) — P. Pommez, M. Lafaix,
    P. Delorme and V. Lorenzelli
Vibrational Analysis — Spectrochim. Acta
    26A, 1503 (1970) — J.H.S. Green
Fluorescence — J. Phys. Chem. 75, 3214
    (1971) — K. Al-Ani and D. Phillips
Calculation — J. Mol. Struct. 8, 319
    (1971) — P.C. Mishra, S.N. Thakur and
    D.K. Rai
Vibrational Analysis — Spectrochim. Acta
    28A, 55 (1972) — L. Verdonck and G.P.
    Van der Kelen
Fluorescence — Chem. Phys. Lett. 14, 404
    (1972) — G.M. Brewer and E.K.C. Lee
Fluorescence — J. Phys. Chem. 78, 7 (1973)—
    M.G. Rockley and D. Phillips

Fluorescence — Chem. Phys. 7, 41 (1975) —
    R.G. Brown, M.G. Rockley and D. Phillips

1-Fluoro-2,4,5-Trichlorobenzene (C_6H_2FCl_3)

$$1\text{-Fluoro-2,4,5-Trichlorobenzene} \quad (C_6H_2FCl_3)$$

Vibrational Analysis — Spectrochim. Acta
    26A, 849 (1970) — R.A. Nyquist

1,2,3-Fluoroxylene (C_6H_3F(CH_3)_2)

$$1,2,3\text{-Fluoroxylene} \quad (C_6H_3F(CH_3)_2)$$

IR Spectrum — Proc. Indian Acad. Sci. A50,
    51 (1959) — M.R. Padhye and T.S.
    Varadarajan
UV Absorption in Vapour — Curr. Sci. 29,
    129 (1960) — M.R. Padhye and T.S.
    Varadarajan
UV Absorption in Vapour — Indian J. Pure
    Appl. Phys. 1, 170 (1963) — M.R.
    Padhye and T.S. Varadarajan
Calculation — Indian J. Pure Appl. Phys.
    1, 25 (1963) — M.R. Padhye and T.S.
    Varadarajan

1,2,4-Fluoroxylene (C_6H_3F(CH_3)_2)

$$1,2,4\text{-Fluoroxylene} \quad (C_6H_3F(CH_3)_2)$$

UV Absorption in Vapour — Curr. Sci. 29,
    129 (1960) — M.R. Padhye and T.S.
    Varadarajan
Calculation — Indian J. Pure Appl. Phys.
    1, 25 (1963) — M.R. Padhye and T.S.
    Varadarajan
UV Absorption in Vapour — Indian J. Pure
    Appl. Phys. 1, 170 (1963) — M.R. Padhye
    and T.S. Varadarajan

1,2,5-Fluoroxylene (C_6H_3F(CH_3)_2)

$$1,2,5\text{-Fluoroxylene} \quad (C_6H_3F(CH_3)_2)$$

IR Spectrum — Proc. Indian Acad. Sci. A50,
    51 (1959) — M.R. Padhye and T.S.
    Varadarajan
UV Absorption in Vapour — Curr. Sci. 29,
    129 (1960) — M.R. Padhye and T.S.
    Varadarajan
Calculation — Indian J. Pure Appl. Phys.
    1, 25 (1963) — M.R. Padhye and T.S.
    Varadarajan
UV Absorption in Vapour — Indian J. Pure
    Appl. Phys. 1, 170 (1963) — M.R.
    Padhye and T.S. Varadarajan

1,2,6-Fluoroxylene ($C_6H_3F(CH_3)_2$)

IR Spectrum — Proc. Indian Acad. Sci. $\underline{A50}$, 51 (1959) — M.R. Padhye and T.S. Varadarajan

UV Absorption in Vapour — Curr. Sci. $\underline{29}$, 129 (1960) — M.R. Padhye and T.S. Varadarajan

Calculation — Indian J. Pure Appl. Phys. $\underline{1}$, 25 (1963) — M.R. Padhye and T.S. Varadarajan

UV Absorption in Vapour — Indian J. Pure Appl. Phys. $\underline{1}$, 170 (1963) — M.R. Padhye and T.S. Varadarajan

Vibrational Analysis — Spectrochim. Acta $\underline{27A}$, 793 (1971) — J.H.S. Green, D.J. Harrison and W. Kynaston

1,3,4-Fluoroxylene ($C_6H_3F(CH_3)_2$)

IR Spectrum — Proc. Indian Acad. Sci. $\underline{A50}$, 51 (1959) — M.R. Padhye and T.S. Varadarajan

UV Absorption in Vapour — Curr. Sci. $\underline{29}$, 129 (1960) — M.R. Padhye and T.S. Varadarajan

Calculation — Indian J. Pure Appl. Phys. $\underline{1}$, 25 (1963) — M.R. Padhye and T.S. Varadarajan

UV Absorption in Vapour — Indian J. Pure Appl. Phys. $\underline{1}$, 170 (1963) — M.R. Padhye and T.S. Varadarajan

Vibrational Analysis — Spectrochim. Acta $\underline{27A}$, 807 (1971) — J.H.S. Green, D.J. Harrison and W. Kynaston

1,3,5-Fluoroxylene ($C_6H_3F(CH_3)_2$)

Vibrational Analysis — Spectrochim. Acta $\underline{27A}$, 793 (1971) — J.H.S. Green, D.J. Harrison and W. Kynaston

o-Formylpyridine (C_5H_4NCHO)

Vibrational Analysis — Spectrochim. Acta $\underline{33A}$, 75 (1977) — J.H.S. Green and D.J. Harrison

m-Formylpyridine (C_5H_4NCHO)

Vibrational Analysis — Spectrochim. Acta $\underline{33A}$, 75 (1977) — J.H.S. Green and D.J. Harrison

p-Formylpyridine (C_5H_4NCHO)

Vibrational Analysis — Spectrochim. Acta $\underline{33A}$, 75 (1977) — J.H.S. Green and D.J. Harrison

Guaicol ($C_6H_4OHOCH_3$)

See o-Methoxyphenol

Hexabromobenzene (C_6Br_6)

Raman — C.R. Acad. Sci.(Paris) $\underline{212}$, 485 (1941) — R. Pajeau

IR Spectrum — J. Chim. Phys. $\underline{64}$, 591 (1967) — P. Delorme, F. Denisselle and V. Lorenzelli

Vibrational Analysis — Spectrochim. Acta $\underline{26A}$, 2261 (1970) — S. Abramowitz and I.W. Levine

Vibrational Analysis — Spectrochim. Acta $\underline{33A}$, 921 (1977) — M. Suzuki

Hexachlorobenzene (C_6Cl_6)

Raman — J. Chem. Phys. $\underline{2}$, 119 (1934) — J.W. Murray and D.H. Andrews

Vibrational Analysis — J. Chem. Phys. $\underline{30}$, 868 (1959) — O. Schnepp and R. Kopelman

Vibrational Analysis — Indian J. Phys. $\underline{34}$, 554 (1960) — K.K. Deb and S.B. Banerjee

Raman — Bull. Chem. Soc. Jap. $\underline{35}$, 322 (1962) — S. Saeki

Vibrational Analysis — Spectrochim. Acta $\underline{19}$, 1739 (1963) — J.R. Scherer and J.C. Evans

IR Spectrum — J. Chim. Phys. $\underline{64}$, 591 (1967) — P. Delorme, F. Denisselle and V. Lorenzelli

Vibrational Analysis — Spectrochim. Acta $\underline{26A}$, 2261 (1970) — S. Abramowitz and I.W. Levine

Vibrational Analysis (Crystal) — Spectrochim. Acta $\underline{27A}$, 637 (1971) — J.B. Bates, D.M. Thomas, A. Bandy and E.R. Lippincott

Hexafluorobenzene (C_6F_6)

Vibrational Analysis — J. Chem. Phys. $\underline{25}$, 182 (1956) — L. Delbouille

Vibrational Analysis — Bull. Acad. Roy. Belgique $\underline{44}$, 971 (1958) — L. Delbouille

IR Spectrum — Trans. Faraday Soc. $\underline{55}$, 369
(1959) — D. Steele and D.H. Whiffen
Calculation — Trans. Faraday Soc. $\underline{56}$, 5
(1960) — D. Steele and D.H. Whiffen
Calculation — Spectrochim. Acta $\underline{19}$, 1947
(1963) — D.A. Long and D. Steele
UV Absorption in Vapour — J. Chem. Phys.
$\underline{39}$, 1253 (1963) — S.H. Bauer and C.F.
Alen
UV Absorption Spectrum — J. Quart. Spectr-
osc. Radiat. Transfer. $\underline{4}$, 819 (1964) —
M. Ballester, J. Palan and J. Riera
Vibrational Analysis — Spectrochim. Acta
$\underline{22}$, 695 (1966) — J.J. Hyams, E.R.
Lippincott and R.T. Bailey
IR Spectrum — Spectrochim. Acta $\underline{22}$, 1723
(1966) — W.B. Pearson, D.A. Olsen and
J.M. Foremwait
Fluorescence — J. Chem. Phys. $\underline{46}$, 4679
(1967) — D. Phillips
Vibrational Analysis — J. Chim. Phys. $\underline{64}$,
591 (1967) — P. Delorme, F. Denisselle
and V. Lorenzelli
IR Spectrum — J. Mol. Spectrosc. $\underline{32}$, 265
(1969) — D. Steele and A.W. Wheatley
Fluorescence — Chem. Phys. Lett. $\underline{13}$, 140
(1972) — G.L. Loper and E.K.C. Lee
Calculation — J. Mol. Spectrosc. $\underline{48}$, 446
(1973) — V.J. Eaton and D. Steele
Vibrational Analysis — J. Chem. Thermodyn-
amics $\underline{8}$, 529 (1976) — J.H.S. Green and
D.J. Harrison

Hexaiodobenzene (C_6I_6)

IR Spectrum — J. Chim. Phys. $\underline{64}$, 591
(1967) — P. Delorme, F. Denisselle
and V. Lorenzelli
Vibrational Analysis — Spectrochim. Acta
$\underline{26A}$, 2261 (1970) — S. Abramowitz and
I.W. Levine
Vibrational Analysis — Spectrochim. Acta
$\underline{33A}$, 921 (1977) — M. Suzuki

Hexamethylbenzene $(C_6(CH_3)_6)$

UV Absorption Spectrum (Crystal) — Z. Elek-
trochem. $\underline{49}$, 372 (1943) — G. Scheibe,
St. Hartwig and R. Muller-Munchen
IR Spectrum — Proc. Roy. Soc. $\underline{A211}$, 168
(1953) — J. Mann and H.W. Thompson
UV Absorption Spectrum (Crystal) — J. Chem.
Phys. $\underline{23}$, 1146 (1955) — R.C. Nelsen
and W.T. Simpson
IR Spectrum — J. Chem. Soc. 3497 (1955) —
R.R. Randle and D.H. Whiffen

UV Absorption in Vapour — J. Chem. Phys.
$\underline{26}$, 83 (1957) — O. Schnepp and D.S.
McClure
UV Absorption Spectrum (Crystal) — Z. Natur-
forsch. $\underline{13A}$, 336 (1958) — H.C. Wolf
IR Spectrum — Spectrochim. Acta $\underline{22}$, 1
(1966) — R.C. Leech, B.D. Powell and
N. Sheppard
UV Absorption Spectrum (Crystal) — J. Mol.
Spectrosc. $\underline{22}$, 452 (1967) — C. Strem-
menoz and C. Zauli

Hydroquinone $(C_6H_4(OH)_2)$

See   1,4-Dihydroxybenzene

Hydroxyanisole $(C_6H_4OHOCH_3)$

See   Methoxyphenol

2-Hydroxyanthraquinone $(C_{14}H_7O_2OH)$

UV Emission in Vapour — Indian J. Pure
Appl. Phys. $\underline{7}$, 713 (1969) — G.D. Baruah,
K.P.R. Nair and B.B. Lal

o-Hydroxybenzaldehyde (C_6H_4CHOOH)

Raman — Z. Phys. Chem. $\underline{38B}$, 119 (1939) — L.
Kahovec and K.W.F. Kohlrausch
UV Absorption in Different States — J. Chem.
Soc. 1347 (1940) — R.A. Mortin and A.L.
Stubbs
Vibrational Analysis — Nature $\underline{166}$, 474
(1950) — A.E. Martin
UV Absorption in Solution — Can. J. Chem.
$\underline{38}$, 1837 (1960) — J.C. Dearden and
W.F. Forbes
UV Absorption in Vapour — Indian J. Phys.
$\underline{34}$, 196 (1960) — I.A. Rao and V.R. Rao
Vibrational Analysis — J. Chem. Soc. 3372
(1961) — C.J.W. Brooks and J.F. Morman
UV Absorption in Vapour — J. Sci. Ind. Res.
(India) $\underline{20B}$, 523 (1961) — I.A. Rao
IR Spectrum — Curr. Sci. $\underline{36}$, 630 (1967) —
O.N. Singh, M.P. Srivastava and I.S.
Singh
Absorption Spectrum — J. Chem. Phys. $\underline{51}$,
1856 (1969) — D.B. Siano and D.F.
Metzler
Absorption Spectrum — Photochem. Photobiol.
$\underline{12}$, 297 (1970) — M. Arrio-Dupont
Absorption in Different States — J. Mol.
Spectrosc. $\underline{53}$, 140 (1974) — C.J.
Seliskar

m—Hydroxybenzaldehyde (C_6H_4OHCHO)

Raman — Z. Phys. Chem. <u>38</u>, 119 (1939) —
K.W.F. Kohlrausch
UV Absorption in Different States — J. Chem.
Soc. 1347 (1940) — R.A. Mortin and
A.L. Stubbs
UV Absorption in Solution — Can. J. Chem.
<u>38</u>, 1837 (1960) — J.C. Dearden and W.F.
Forbes
UV Absorption in Vapour — Indian J. Phys.
<u>34</u>, 196 (1960) — I.A. Rao and V.R. Rao
UV Absorption in Vapour — J. Sci. Ind. Res.
(India) <u>20B</u>, 523 (1961) — I.A. Rao
IR Spectrum — J. Sci. Res. BHU (India) <u>18</u>,
200 (1967) — G.D. Baruah, O.N. Singh,
M.P. Srivastava and I.S. Singh
UV Emission in Vapour — Indian J. Pure
Appl. Phys. <u>7</u>, 352 (1969) — G.D. Baruah,
O.N. Singh and R.S. Singh

p—Hydroxybenzaldehyde (C_6H_4OHCHO)

Raman — Z. Phys. Chem. <u>38</u>, 119 (1939) —
K.W.F. Kohlrausch
UV Absorption in Different States — J. Chem.
Soc. 1347 (1940) — R.A. Mortin and
A.L. Stubbs
UV Absorption in Vapour — Indian J. Phys.
<u>34</u>, 196 (1960) — I.A. Rao and V.R. Rao
UV Absorption in Vapour — J. Sci. Ind. Res.
(India) <u>20B</u>, 523 (1961) — I.A. Rao
Vibrational Analysis — Appl. Spectrosc. <u>16</u>,
32 (1962) — R.J. Jackobsen and E.J.
Brewer
IR Spectrum — J.Sci. Res. BHU (India) <u>18</u>,
200 (1967) — G.D. Baruah, M.P. Srivas-
tava, O.N. Singh and I.S. Singh
UV Emission in Vapour — Indian J. Pure Appl.
Phys. <u>7</u>, 352 (1969) — G.D. Baruah,
O.N. Singh and R.S. Singh

p—Hydroxybenzonitrile (C_6H_4OHCN)

Vibrational Analysis — Z. Elektrochem. <u>64</u>,
1218 (1960) — S. Weckherlin and W.
Luttke
Vibrational Analysis — Appl. Spectrosc. <u>16</u>,
32 (1962) — R.J. Jackobsen and E.J.
Brewer
Vibrational Analysis — Spectrochim. Acta
<u>21</u>, 45 (1965) — H.W. Wilson and J.E.
Bloor

2—Hydroxy—3—Methoxybenzaldehyde ($C_6H_3OHOCH_3CHO$)

IR Spectrum — Anal. Chem. <u>27</u>, 2 (1955) —
S. Pinchas
IR Spectrum — Anal. Chem. <u>29</u>, 334 (1957) —
S. Pinchas

3—Hydroxy—4—Methoxybenzaldehyde ($C_6H_3OHOCH_3CHO$)

IR Spectrum — Spectrochim. Acta <u>10</u>, 21
(1957) — M.S.C. Flett

o—Hydroxypyridine (C_5H_4NOH)

IR Spectrum — J. Chem. Soc. 4340 (1955) —
J.A. Gibson, W. Kynaston and A.S.
Lindsey
IR Spectrum — J. Chem. Soc. 4874 (1957) —
S.F. Mason
IR Spectrum — J. Chem. Soc. 2947 (1960) —
A.R. Katritzky and R.A. Jones
UV Absorption in Vapour — Indian J. Pure
Appl. Phys. <u>13</u>, 417 (1975) — R.S.
Tripathi and B.R. Pandey
IR Spectrum — Indian J. Pure Appl. Phys.
<u>13</u>, 416 (1975) — R.S. Tripathi and
B.R. Pandey

m—Hydroxypyridine (C_5H_4NOH)

IR Spectrum — Indian J. Pure Appl. Phys.
<u>13</u>, 416 (1975) — R.S. Tripathi and
B.R. Pandey

p—Hydroxypyridine (C_5H_4NOH)

Phosphorescence — Spectrochim. Acta <u>32A</u>,
1659 (1976) — S. Hotchandani and
A.C. Testa

Hydroxytoluene ($C_6H_4CH_3OH$)

See Methylphenol

o—Iodoaniline ($C_6H_4NH_2I$)

IR Spectrum — Spectrochim. Acta <u>29A</u>,
1555 (1973) — J.M. Briody

p-Iodoaniline $(C_6H_4NH_2I)$

Calculation — J. Mol. Struct. <u>14</u>, 61
(1972) — L. Smetankine and J. Etchepare
IR Spectrum — Spectrochim. Acta <u>29A</u>, 1555
(1973) — J.M. Briody

p-Iodoanisole $(C_6H_4OCH_3I)$

UV Absorption in Solution — Spectrochim.
Acta <u>17</u>, 545 (1961) — E. Spinner
Vibrational Analysis — Spectrochim. Acta
<u>23A</u>, 1111 (1967) — M. Horak, E.R.
Lippincott and R.K. Khanna
IR Spectrum — Spectrochim. Acta <u>29A</u>, 1555
(1973) — J.M. Briody

o-Iodobenzaldehyde (C_6H_4CHOI)

UV Absorption in Solution — Can. J. Chem.
<u>36</u>, 1362 (1958) — J.C. Dearden and
W.F. Forbes
Vibrational Analysis — Indian J. Pure Appl.
Phys. <u>11</u>, 615 (1973) — B.B. Lal, M.P.
Srivastava and I.S. Singh

m-Iodobenzaldehyde (C_6H_4CHOI)

Vibrational Analysis — Anal. Chem. <u>23</u>, 334
(1957) — S. Pinchas
UV Absorption in Solution — Can. J. Chem.
<u>36</u>, 1362 (1958) — J.C. Dearden and
W.F. Forbes
Raman — Spectrochim. Acta <u>23A</u>, 462 (1967) —
F.B. Brown
Vibrational Analysis — Indian J. Pure Appl.
Phys. <u>11</u>, 615 (1973) — B.B. Lal, M.P.
Srivastava and I.S. Singh

p-Iodobenzaldehyde (C_6H_4CHOI)

UV Absorption in Solution — Can. J. Chem.
<u>36</u>, 1362 (1958) — J.C. Dearden and
W.F. Forbes
Vibrational Analysis — Indian J. Pure
Appl. Phys. <u>11</u>, 615 (1973) — B.B. Lal,
M.P. Srivastava and I.S. Singh

Iodobenzene (C_6H_5I)

Vibrational Analysis — Proc. Indian Acad.
Sci. <u>A15</u>, 401 (1942) — C.S. Venkateshwaran and N.S. Pandya

UV Absorption in Vapour — Curr. Sci.
(1950) — K. Sreeramamurty and
K.R. Rao
IR Spectrum — J. Chem. Soc. 1350 (1956) —
D.H. Whiffen
IR Spectrum — Indian J. Phys. <u>38</u>, 610
(1964) — S.C. Sirkar, D.K. Mukherjee
and P.K. Bishnui
IR Spectrum — Spectrochim. Acta <u>22</u>, 501
(1966) — W.R. McWhinnie and R.C. Poller
Vibrational Analysis — Proc. Roy. Soc.
<u>A298</u>, 51 (1967) — P.R. Griffiths and
H.W. Thompson
Raman — J. Mol. Struct. <u>5</u>, 477 (1970) —
N.T. McDevitt and W.G. Fateley
Raman — J. Mol. Spectrosc. <u>39</u>, 73 (1971) —
C.T. Mennely, C.Y. She and D.F.
Edwards

Iodobenzene-D5 (C_6D_5I)

Vibrational Analysis — Spectrochim. Acta
<u>22</u>, 737 (1966) — T.R. Nanney and E.R.
Lippincott

o-Iodobenzoic acid $(C_6H_4HCO_2I)$

Vibrational Analysis — Spectrochim. Acta
<u>33A</u>, 575 (1977) — J.H.S. Green

m-Iodobenzoic acid $(C_6H_4HCO_2I)$

Vibrational Analysis — Spectrochim. Acta
<u>33A</u>, 575 (1977) — J.H.S. Green

p-Iodobenzoic acid $(C_6H_4HCO_2I)$

Vibrational Analysis — Spectrochim. Acta
<u>33A</u>, 575 (1977) — J.H.S. Green

α-Iodonaphthalene $(C_{10}H_7I)$

UV Absorption in Solution — J. Chem. Soc.
304 (1954) — J. Ferguson
Phosphorescence — J. Chem. Soc. 3160
(1954) — J. Ferguson, T. Iredale and
J.A. Taylor
Vibrational Analysis — Spectrochim. Acta
<u>34A</u>, 985 (1978) — S.N. Singh, H.S.
Bhatti and R.D. Singh

β-Iodonaphthalene ($C_{10}H_7I$)

UV Absorption in Solution - J. Chem. Soc.
304 (1954) - J. Ferguson
Phosphorescence - J. Chem. Soc. 3160
(1954) - J. Ferguson, T. Iredale and
J.A. Taylor

o-Iodonitrobenzene ($C_6H_4INO_2$)

Vibrational Analysis - J. Chim. Phys. 62,
347 (1965) - M. Brigodiot and J.M.
Lebas
Vibrational Analysis - Spectrochim. Acta
26A, 1925 (1970) - J.H.S. Green and
D.J. Harrison
IR Spectrum - Spectrochim. Acta 29A, 1555
(1973) - J.M. Briody

m-Iodonitrobenzene ($C_6H_4INO_2$)

Vibrational Analysis - J. Chim. Phys. 63,
552 (1966) - C. Garrigou-Lagrange,
M. Chehata and J. Lascombe
Vibrational Analysis - Spectrochim. Acta
26A, 1925 (1970) - J.H.S. Green and
D.J. Harrison

p-Iodonitrobenzene ($C_6H_4INO_2$)

IR Spectrum - J. Amer. Chem. Soc. 78, 4225
(1956) - R.D. Kross and V.A. Fassel
Vibrational Analysis - J. Chim. Phys. 64,
1450 (1967) - P. Pommez, M. Lafaix, P.
Delorme and V. Lorenzelli
Vibrational Analysis - Proc. Roy. Soc.
A298, 51 (1967) - P.R. Griffiths and
H.W. Thompson
Vibrational Analysis - Spectrochim. Acta
26A, 1925 (1970) - J.H.S. Green and
D.J. Harrison
Calculation - J. Mol. Struct. 14, 61
(1972) - L. Smetankine and J. Etchepare
IR Spectrum - Spectrochim. Acta 29A, 1555
(1973) - J.M. Briody

o-Iodophenol (C_6H_4IOH)

Vibrational Analysis - Z. Elektrochem. 59,
866 (1955) - R. Mecke and G. Rossmy
Vibrational Analysis - Spectrochim. Acta
27A, 2199 (1971) - J.H.S. Green, D.J.
Harrison and W. Kynaston

o-Iodophenol-OD (C_6H_4IOD)

Vibrational Analysis - Spectrochim. Acta
27A, 2199 (1971) - J.H.S. Green, D.J.
Harrison and W. Kynaston

m-Iodophenol (C_6H_4IOH)

Vibrational Analysis - Spectrochim. Acta
27A, 2199 (1971) - J.H.S. Green, D.J.
Harrison and W. Kynaston

p-Iodophenol (C_6H_4IOH)

Vibrational Analysis - Spectrochim. Acta
27A, 2199 (1971) - J.H.S. Green, D.J.
Harrison and W. Kynaston

o-Iodopyrimidine ($C_4H_3N_2I$)

Vibrational Analysis - Spectrochim. Acta
33A, 189 (1977) - E. Allenstein, P.
Kiemle, J. Weidlein and W. Podszun

o-Iodotoluene ($C_6H_4CH_3I$)

Raman - Monatsh. Chem. 76, 1 (1946) - E.
Herz
Vibrational Analysis - Spectrochim. Acta
26A, 1913 (1970) - J.H.S. Green
IR Spectrum - Spectrochim. Acta 29A, 1555
(1973) - J.M. Briody

m-Iodotoluene ($C_6H_4CH_3I$)

Vibrational Analysis - Spectrochim. Acta
26A, 1523 (1970) - J.H.S. Green

p-Iodotoluene ($C_6H_4CH_3I$)

Vibrational Analysis - Spectrochim. Acta
12, 305 (1958) - C. Garrigou-Lagrange,
J.M. Lebas and M.L. Josien
Vibrational Analysis - Spectrochim. Acta
13, 225 (1959) - J.M. Lebas, C. Garrigou-
Lagrange and M.L. Josien
Vibrational Analysis - Spectrochim. Acta
20, 1343 (1964) - E.F. Mooney
Vibrational Analysis - Spectrochim. Acta
26A, 1503 (1970) - J.H.S. Green
IR Spectrum - Spectrochim. Acta 29A, 1555
(1973) - J.M. Briody

Isophthaldehyde ($C_6H_4(CHO)_2$)

IR Spectrum — Indian J. Pure Appl. Phys. 10, 570 (1972) — M.P. Srivastava, B.B. Lal and I.S. Singh

p-Isopropylbenzaldehyde ($C_6H_4CHOCH(CH_3)_2$)

UV Absorption in Different States — J. Chem. Soc. 2482 (1914) — J.E. Purvis
UV Absorption in Solution — J. Chem. Soc. 1408 (1938) — R.G. Cooke and A.K. Macbeth
UV Absorption in Vapour — Indian J. Pure Appl. Phys. 7, 583 (1969) — H.G. Srinivasacharya

Isoquinoline (C_9H_7N)

Raman — Indian J. Phys. 10, 23 (1937) — K. Jatkar
UV Absorption in Solution — J. Amer. Chem. Soc. 68, 2181 (1946) — G.W. Ewing and E.A. Steck
Vibrational Analysis — J. Prakt. Chem. 5, 242 (1957) — H. Luther, D. Mootz and F. Radwitz
Calculation — Bull. Chem. Soc. Jap. 31, 459 (1958) — N. Mataza
UV Absorption in Solution — Z. Elektrochem. 65, 61 (1961) — H. Zimmerman and N. Joop
Vibrational Analysis — Indian J. Phys. 35, 535 (1961) — K.K. Deb
Calculation — J. Chem. Phys. 36, 1948 (1962) — M.A. El-Sayed
Calculation — J. Chem. Phys. 36, 1993 (1962) — M.A. El-Sayed
Vibrational Analysis — Indian J. Phys. 36, 557 (1962) — K.K. Deb
UV Absorption in Solution — Spectrochim. Acta 18, 1441 (1962) — G. Coppens, C. Gillet, J. Nascilski and E.V. Donckt
Vibrational Analysis — J. Mol. Spectrosc. 34, 56 (1970) — S.C. Wait, Jr and J.C. McNersey
IR Spectrum — Appl. Spectrosc. 24, 344 (1970) — R. Amni Amma, S.N. Thakur and K.P.R. Nair
UV Absorption in Vapour — Appl. Spectrosc. 24, 344 (1970) — R. Amni Amma, S.N. Thakur and K.P.R. Nair
Calculation — J. Mol. Spectrosc. 44, 118 (1972) — H. Baba and Y. Yamazuki

UV Absorption in Vapour — Indian J. Pure Appl. Phys. 11, 73 (1973) — M.A. Shashidhar
UV Absorption Spectrum — J. Mol. Spectrosc. 47, 189 (1973) — U.T. Kreibich and Urs.P. Wild
Calculation — J. Mol. Spectrosc. 50, 457 (1974) — J.E. Ridley and M.C. Zerner

Mesidine ($C_6H_2NH_2(CH_3)_2$)

UV Absorption in Solution — Bull. Soc. Chim. Fr. 45, 134 (1947) — P. Grammaticakis
IR Spectrum — Proc. Natl. Dakota Akad. Sci. 12, 33 (1958) — E.J. Kaatz and E.J. O'Reilly
Vibrational Analysis — Indian J. Pure Appl. Phys. 8, 678 (1970) — G.D. Baruah, K. Singh and D.K. Rai

Mesitylene ($C_6H_3(CH_3)_3$)

See 1,3,5-Trimethylbenzene

o-Methoxybenzaldehyde ($C_6H_4OCH_3CHO$)

Raman — Z. Phys. Chem. 38B, 119 (1937) — L. Kahovec and K.W.F. Kohlrausch
UV Absorption in Different States — J. Chem. Soc. 1347 (1940) — R.A. Morton and A.L. Stubbs
UV Absorption in Solution — Bull. Soc. Chim. Fr. 51, 821 (1953) — I.P. Grammaticakis
IR Spectrum — Anal. Chem. 27, 2 (1955) — S. Pinchas
IR Spectrum — J. Chem. Soc. 3670 (1959) — A.R. Katritzky and R.A. Jones
Vibrational Analysis — J. Amer. Chem. Soc. 81, 3933 (1959) — D. Heinert and A.E. Martell
Raman — Bull. Soc. Chim. Belges 68, 643 (1959) — G. Michel
UV Absorption in Solution — Can. J. Chem. 38, 1837 (1960) — J.C. Dearden and W.F. Forbes
IR Spectrum — J. Chem. Soc. 3372 (1961) — C.J.W. Brooks and J.F. Norman
IR Spectrum — Bull. Chem. Soc. Jap. 36, 1020 (1963) — H. Minato
UV Absorption in Vapour — Indian J. Pure Appl. Phys. 3, 405 (1965) — R.M.P. Jaiswal
UV Absorption in Vapour — Indian J. Pure Appl. Phys. 3, 452 (1965) — C.P.D. Dwivedi

Vibrational Analysis — Curr. Sci. <u>37</u>, 100
(1968) — M.P. Srivastava, O.N. Singh
and I.S. Singh
Vibrational Analysis — Indian J. Pure Appl.
Phys. <u>6</u>, 440 (1968) — C.P.D. Dwivedi
IR Spectrum — Indian J. Pure Appl. Phys.
<u>7</u>, 504 (1969) — M.P. Srivastava, O.N.
Singh and I.S. Singh
UV Absorption in Vapour — Indian J. Pure
Appl. Phys. <u>7</u>, 410 (1969) — C.P.D.
Dwivedi

m—Methoxybenzaldehyde $(C_6H_4OCH_3CHO)$

Raman — Z. Phys. Chem. <u>38B</u>, 119 (1937) —
L. Kahovec and K.W.F. Kohlrausch
UV Absorption in Different States — J.
Chem. Soc. 1347 (1940) — R.A. Morton
and A.L. Stubbs
UV Absorption in Solution — Bull. Soc.
Chim. Fr. <u>51</u>, 821 (1953) — I.P.
Grammaticakis
Raman — Bull. Soc. Chim. Belges <u>68</u>, 643
(1959) — G. Michel
IR Spectrum — J. Chem. Soc. 3372 (1961) —
C.J.W. Brooks and J.F. Norman
UV Absorption in Vapour — Indian J. Pure
Appl. Phys. <u>3</u>, 452 (1965) — C.P.D.
Dwivedi
UV Absorption in Vapour — Indian J. Pure
Appl. Phys. <u>3</u>, 405 (1965) — R.M.P.
Jaiswal
Vibrational Analysis — Curr. Sci. <u>37</u>, 100
(1968) — M.P. Srivastava, O.N. Singh
and I.S. Singh
Vibrational Analysis — Indian J. Pure
Appl. Phys. <u>6</u>, 440 (1968) — C.P.D.
Dwivedi
UV Absorption in Vapour — Indian J. Pure
Appl. Phys. <u>7</u>, 410 (1969) — C.P.D.
Dwivedi
IR Spectrum — Indian J. Pure Appl. Phys.
<u>7</u>, 504 (1969) — M.P. Srivastava, O.N.
Singh and I.S. Singh
Vibrational Analysis — J. Sci. Res. BHU
(India) <u>20</u>, 158 (1969) — M.P. Srivast-
ava, O.N. Singh and I.S. Singh

p—Methoxybenzaldehyde $(C_6H_4OCH_3CHO)$

Raman — Z. Phys. Chem. <u>38B</u>, 119 (1937) — L.
Kahovec and K.W.F. Kohlrausch
UV Absorption in Different States — J. Chem.
Soc. 1347 (1940) — R.A. Mortin and A.L.
Stubbs
UV Absorption in Solution — Bull. Soc. Chim.
Fr. <u>51</u>, 821 (1953) — I.P. Grammaticakis

IR Spectrum — Anal. Chem. <u>27</u>, 2 (1955) —
S. Pinchas
UV Absorption in Vapour — J. Sci. Ind.
Res.(India) <u>16B</u>, 230 (1957) — V.
Suryanarayana and V.R. Rao
Raman — Bull. Soc. Chim. Belges <u>68</u>, 643
(1959) — G. Michel
UV Absorption — Spectrochim. Acta <u>17</u>,
545 (1961) — E. Spinner
UV Absorption in Vapour — Indian J. Pure
Appl. Phys. <u>3</u>, 405 (1965) — R.M.P.
Jaiswal
UV Absorption in Vapour — Indian J. Pure
Appl. Phys. <u>3</u>, 452 (1965) — C.P.D.
Dwivedi
Vibrational Analysis — Curr. Sci. <u>37</u>,
100 (1968) — M.P. Srivastava, O.N.
Singh and I.S. Singh
Vibrational Analysis — Indian J. Pure
Appl. Phys. <u>6</u>, 440 (1968) — C.P.D.
Dwivedi
IR Spectrum — Indian J. Pure Appl. Phys.
<u>7</u>, 504 (1969) — M.P. Srivastava, O.N.
Singh and I.S. Singh
UV Absorption in Vapour — Indian J. Pure
Appl. Phys. <u>7</u>, 410 (1969) — C.P.D.
Dwivedi

Methoxybenzene $(C_6H_5OCH_3)$

See   Anisole

o—Methoxybenzonitrile $(C_6H_4CNOCH_3)$

UV Absorption in Solution — J. Mol.
Spectrosc. <u>55</u>, 163 (1975) — Y.H. Lui
and S.P. McGlynn
Fluorescence — J. Mol. Spectrosc. <u>55</u>,
163 (1975) — Y.H. Lui and S.P.
McGlynn

m—Methoxybenzonitrile $(C_6H_4CNOCH_3)$

UV Absorption in Solution — J. Mol.
Spectrosc. <u>55</u>, 163 (1975) — Y.H. Lui
and S.P. McGlynn
Fluorescence — J. Mol. Spectrosc. <u>55</u>,
163 (1975) — Y.H. Lui and S.P.
McGlynn

p—Methoxybenzonitrile $(C_6H_4CNOCH_3)$

Vibrational Analysis — Z. Elektrochem.
<u>64</u>, 1228 (1960) — S. Weckherlin
and W. Luttke

Vibrational Analysis — Spectrochim. Acta
21, 45 (1965) — H.W. Wilson and J.E.
Bloor
UV Absorption in Solution — J. Mol.
Spectrosc. 55, 163 (1975) — Y.H. Lui
and S.P. McGlynn
Fluorescence — J. Mol. Spectrosc. 55, 163
(1975) — Y.H. Lui and S.P. McGlynn

α—Methoxynaphthalene ($C_{10}H_7OCH_3$)

UV Absorption in Vapour — Spectrochim. Acta
30A, 277 (1974) — D. Marjit and S.B.
Banerjee
IR Spectrum — Indian J. Pure Appl. Phys.
12, 727 (1974) — D. Marjit and S.B.
Banerjee

2—Methoxy—4—Nitroaniline ($C_6H_3NH_2NO_2OCH_3$)

IR Spectrum — Aust. J. Chem. 23, 947
(1970) — L.K. Dyall

o—Methoxyphenol ($C_6H_4OCH_3OH$)

Raman — Monatsh. Chem. 66, 285 (1935) —
A.W. Reitz and G.P. Ypsilanti
Raman — Monatsh. Chem. 76, 1 (1946) — E.
Herz
UV Absorption in Vapour — J. Sci. Ind. Res.
(India) 15B, 260 (1956) — V. Suryanara-
yana and V.R. Rao
UV Absorption in Different States — Indian
J. Phys. 30, 553 (1956) — S.K. Sen
Vibrational Analysis — Suomen Kemistilehti
A32, 193 (1959) — J.J. Linderberg and
J. Kenttamaa
IR Spectrum — Indian J. Phys. 42, 325
(1968) — D.K. Mukharjee and S.B.
Banerjee
Vibrational Analysis — Israel J. Chem. 8,
777 (1970) — V.N. Verma, K.P.R. Nair
and D.K. Rai
Vibrational Analysis — Spectrochim. Acta
27A, 2199 (1971) — J.H.S. Green, D.J.
Harrison and W. Kynaston

o—Methoxyphenol—OD ($C_6H_4ODOCH_3$)

Vibrational Analysis — Spectrochim. Acta
27A, 2199 (1971) — J.H.S. Green, D.J.
Harrison and W. Kynaston

m—Methoxyphenol ($C_6H_4OCH_3OH$)

Raman — Monatsh. Chem. 66, 299 (1935) —
A.W. Reitz and G.P. Ypsilanti
Vibrational Analysis — Israel J. Chem. 8,
777 (1970) — V.N. Verma, K.P.R. Nair
and D.K. Rai
UV Absorption in Vapour — Indian J. Phys.
44, 212 (1970) — C.G. Rama Rao, B.R.K.
Reddy and P.T. Rao

p—Methoxyphenol ($C_6H_4OCH_3OH$)

Raman — Monatsh. Chem. 66, 299 (1935) —
A.W. Reitz and G.P. Ypsilanti
Raman — Monatsh. Chem. 72, 244 (1939) —
O. Paulsen
UV Absorption in Vapour — J. Sci. Ind. Res.
(India) 15B, 548 (1956) — V. Suryana-
rayana and V.R. Rao
Vibrational Analysis — Appl. Spectrosc. 16,
32 (1962) — R.J. Jackobsen and E.J.
Brewer
Vibrational Analysis — Israel J. Chem. 8,
777 (1970) — V.N. Verma, K.P.R. Nair
and D.K. Rai

o—Methoxypyridine ($C_5H_4NOCH_3$)

UV Absorption in Solution — Spectrochim.
Acta 24A, 207 (1968) — G. Favini, M.
Raimondi and G. Gandolfo

m—Methoxypyridine ($C_5H_4NOCH_3$)

UV Absorption in Solution — Spectrochim.
Acta 24A, 207 (1968) — G. Favini, M.
Raimondi and G. Gandolfo

p—Methoxypyridine ($C_5H_4NOCH_3$)

UV Absorption in Solution — Spectrochim.
Acta 24A, 207 (1968) — G. Favini, M.
Raimondi and G. Gandolfo

p—Methoxystyrene ($C_6H_4CH{:}CH_2OCH_3$)

IR Spectrum — Indian J. Pure Appl. Phys.
8, 725 (1970) — B.J. Ansari

o—Methoxytoluene ($C_6H_4CH_3OCH_3$)

Raman — Monatsh. Chem. <u>76</u>, 1 (1946) — E. Herz

UV Absorption in Vapour — Indian J. Phys. <u>34</u>, 200 (1960) — K.V.K. Rao and V.R. Rao

Vibrational Analysis — Spectrochim. Acta <u>27A</u>, 2199 (1971) — J.H.S. Green, D.J. Harrison and W. Kynaston

m—Methoxytoluene ($C_6H_4CH_3OCH_3$)

UV Absorption in Vapour — Indian J. Phys. <u>34</u>, 200 (1960) — K.V.K. Rao and V.R. Rao

IR Spectrum — Indian J. Pure Appl. Phys. <u>11</u>, 787 (1973) — C.P.D. Dwivedi and S.N. Sharma

p—Methoxytoluene ($C_6H_4CH_3OCH_3$)

UV Absorption in Vapour — Indian J. Phys. <u>34</u>, 200 (1960) — K.V.K. Rao and V.R. Rao

IR Spectrum — Indian J. Pure Appl. Phys. <u>11</u>, 787 (1973) — C.P.D. Dwivedi and S.N. Sharma

N—Methylaniline ($C_6H_5NH:CH_3$)

IR Spectrum — J. Amer. Chem. Soc. <u>47</u>, 2192 (1925) — K.F. Bell

Raman — Z. Phys. <u>70</u>, 131 (1931) — A.S. Ganesan and V.N. Thatte

Raman — Monatsh. Chem. <u>69</u>, 363 (1936) — L. Kahovec and A.W. Reitz

IR Spectrum — C.R. Acad. Sci.(Paris) <u>20B</u>, 852 (1937) — M. Freymann

UV Absorption in Different States — J. Amer. Chem. Soc. <u>72</u>, 1543 (1950) — W.W. Robertson and F.A. Matsen

Vibrational Analysis — Proc. Roy. Soc. <u>A243</u>, 143 (1957) — P.J. Krueger and H.W. Thompson

IR Spectrum — J. Chem. Soc. 843 (1957) — D. Hadzi and M. Skrbljak

IR Spectrum — J. Chem. Soc. 760 (1958) — R.D. Hill and G.D. Meakins

UV Absorption in Vapour — Indian J. Pure Appl. Phys. <u>1</u>, 192 (1963) — C.P.D. Dwivedi and D. Sharma

UV Emission in Vapour — Indian J. Pure Appl. Phys. <u>3</u>, 219 (1965) — S.N. Singh

UV Absorption in Vapour — Indian J. Pure Appl. Phys. <u>7</u>, 186 (1969) — C.P.D. Dwivedi

Vibrational Analysis — Indian J. Pure Appl. Phys. <u>9</u>, 336 (1971) — V.N. Verma, K.P.R. Nair and D.K. Rai

Vibrational Analysis — Indian J. Pure Appl. Phys. <u>16</u>, 454 (1978) — A.K. Ansari and P.K. Verma

o—Methylaniline ($C_6H_4NH_2CH_3$)

IR Spectrum — Spectrochim. Acta <u>16</u>, 677 (1960) — A. Sabatnini and S. Califano

Vibrational Analysis — Proc. Indian Acad. Sci. <u>54</u>, 146 (1961) — P.G. Puranik and K. Venkataramaiah

IR Spectrum — Spectrochim. Acta <u>17</u>, 614 (1961) — K.B. Whetsel

IR Spectrum — Indian J. Phys. <u>36</u>, 457 (1962) — K.C. Medhi, S.B. Banerjee and G.S. Kastha

Raman — J. Mol. Struct. <u>5</u>, 477 (1970) — N.T. McDevitt and W.G. Fateley

UV Absorption in Different States — Indian J. Phys. <u>51B</u>, 392 (1977) — T. Ganguly and S.B. Banerjee

Vibrational Analysis — Indian J. Pure Appl. Phys. <u>16</u>, 454 (1978) — A.K. Ansari and P.K. Verma

m—Methylaniline ($C_6H_4NH_2CH_3$)

IR Spectrum — Spectrochim. Acta <u>16</u>, 677 (1960) — A. Sabatnini and S. Califano

IR Spectrum — Spectrochim. Acta <u>16</u>, 352 (1960) — J.C. Evans and N. Wright

Vibrational Analysis — Proc. Indian Acad. Sci. <u>54</u>, 146 (1961) — P.G. Puranik and K. Venkataramaiah

IR Spectrum — Indian J. Phys. <u>36</u>, 457 (1962) — K.C. Medhi, S.B. Banerjee and G.S. Kastha

Vibrational Analysis — J. Chim. Phys. <u>63</u>, 552 (1966) — C. Garrigou-Lagrange, M. Chehata and J. Lascombe

UV Absorption in Different States — Indian J. Phys. <u>51B</u>, 392 (1977) — T. Ganguly and S.B. Banerjee

Vibrational Analysis — Indian J. Pure Appl. Phys. <u>16</u>, 454 (1978) — A.K. Ansari and P.K. Verma

p—Methylaniline ($C_6H_4NH_2CH_3$)

Vibrational Analysis — Spectrochim. Acta <u>12</u>, 305 (1958) — C. Garrigou-Lagrange, J.M. Lebas and M.L. Josien

Vibrational Analysis — Spectrochim. Acta
15, 225 (1959) — J.M. Lebas, C.
Garrigou-Lagrange and M.L. Josien
IR Spectrum — Spectrochim. Acta 16, 677
(1960) — A. Sabatnini and S. Califano
Vibrational Analysis — Proc. Indian Acad.
Sci. 54, 146 (1961) — P.G. Purabik and
K. Venkataramaiah
IR Spectrum — Indian J. Phys. 36, 457
(1962) — K.C. Medhi, S.B. Banerjee
and G.S. Kastha
IR Spectrum — Spectrochim. Acta 25A, 1423
(1969) — L.K. Dyall
UV Absorption in Different States — Indian
J. Phys. 51B, 392 (1977) — T. Ganguly
and S.B. Banerjee

Methylanisole $(C_6H_4CH_3OCH_3)$

See   Methoxytoluene

1-Methylanthraquinone $(C_{14}H_7O_2CH_3)$

UV Emission in Vapour — Curr. Sci. 36, 483
(1967) — S. Nath Singh and R.S. Singh
Vibrational Analysis — Curr. Sci. 36, 604
(1967) — S. Nath Singh and R.S. Singh

2-Methylanthraquinone $(C_{14}H_7O_2CH_3)$

UV Emission in Vapour — Indian J. Pure
Appl. Phys. 7, 520 (1969) — G.D.
Baruah, K.P.R. Nair and D.K. Rai

o-Methylbezaldehyde $(C_6H_4CHOCH_3)$

UV Absorption in Solution — Spectrochim.
Acta 28A, 1969 (1972) — E.V. Donckt
and C. Vogels
Vibrational Analysis — Spectrochim. Acta
32A, 1265 (1976) — J.H.S. Green and
D.J. Harrison

m-Methylbenzaldehyde $(C_6H_4CHOCH_3)$

UV Absorption in Solution — Spectrochim.
Acta 28A, 1969 (1972) — E.V. Donckt
and C. Vogels
Vibrational Analysis — Spectrochim. Acta
32A, 1265 (1976) — J.H.S. Green and
D.J. Harrison

p-Methylbenzaldehyde $(C_6H_4CHOCH_3)$

UV Absorption in Vapour — J. Sci. Res.
BHU (India) 14, 76 (1963) — N.L.
Singh and R.N. Singh
UV Absorption in Solution — Spectrochim.
Acta 28A, 1969 (1972) — E.V. Donckt
and C. Vogels
Vibrational Analysis — Spectrochim. Acta
32A, 1265 (1976) — J.H.S. Green and
D.J. Harrison

o-Methylbenzoic acid $(C_6H_4CH_3CO_2H)$

Vibrational Analysis — Spectrochim. Acta
33A, 575 (1977) — J.H.S. Green

m-Methylbenzoic acid $(C_6H_4CH_3CO_2H)$

Vibrational Analysis — Spectrochim. Acta
33A, 575 (1977) — J.H.S. Green

p-Methylbenzoic acid $(C_6H_4CH_3CO_2H)$

Vibrational Analysis — Spectrochim. Acta
33A, 575 (1977) — J.H.S. Green

o-Methylbenzonitrile $(C_6H_4CNCH_3)$

Raman — Monatsh. Chem. 63, 427 (1933) —
K.W.F. Kohlrausch and A. Pongratz
Vibrational Analysis — J. Chim. Phys. 62,
347 (1965) — M. Brigodiot and J.M.
Lebas
Vibrational Analysis — Spectrochim. Acta
32A, 1279 (1976) — J.H.S. Green and
D.J. Harrison

m-Methylbenzonitrile $(C_6H_4CNCH_3)$

Raman — Monatsh. Chem. 63, 427 (1933) —
K.W.F. Kohlrausch and A. Pongratz
Vibrational Analysis — Indian J. Pure Appl.
Phys. 14, 419 (1976) — S.P. Sinha and
C.L. Chatterjee
Vibrational Analysis — Spectrochim. Acta
32A, 1279 (1976) — J.H.S. Green and
D.J. Harrison

p-Methylbenzonitrile $(C_6H_4CNCH_3)$

Raman — Monatsh. Chem. 63, 427 (1933) —
K.W.F. Kohlrausch and A. Pongratz

Vibrational Analysis — Spectrochim. Acta
$\underline{21}$, 45 (1965) — H.W. Wilson and J.E.
Bloor

Vibrational Analysis — Indian J. Pure Appl.
Phys. $\underline{14}$, 421 (1976) — S.P. Sinha and
C.L. Chatterjee

Vibrational Analysis — Spectrochim. Acta
$\underline{32A}$, 1279 (1976) — J.H.S. Green and
D.J. Harrison

Methyl-p-benzoquinone $(C_6H_3O_2CH_3)$

See   Toluquinone

m-Methylbenzylbromide $(C_6H_4CH_3CH_2Br)$

Vibrational Analysis — Spectrochim. Acta
$\underline{29A}$, 813 (1973) — L. Verdonck, G.P.
Van der Kelen and Z. Eeckhaut

m-Methylbenzylchloride $(C_6H_4CH_3CH_2Cl)$

Vibrational Analysis — Spectrochim. Acta
$\underline{29A}$, 817 (1973) — L. Verdonck, G.P.
Van der Kelen and Z. Eeckhaut

2-Methyl-1,3-Dichlorobenzene
$(C_6H_3CH_3Cl_2)$

Vibrational Analysis — Spectrochim. Acta
$\underline{27A}$, 793 (1971) — J.H.S. Green, D.J.
Harrison and W. Kynaston

3-Methylisoquinoline $(C_9H_6NCH_3)$

IR Spectrum — Indian J. Pure Appl. Phys.
$\underline{12}$, 532 (1974) — M.A. Shashidhar

UV Absorption in Vapour — Indian J. Pure
Appl. Phys. $\underline{12}$, 532 (1974) — M.A.
Shashidhar

α-Methylnaphthalene $(C_{10}H_7CH_3)$

UV Absorption in Vapour — Z. Phys. Chem.
$\underline{11B}$, 369 (1925) — H.G. De'Laszlo

UV Absorption in Different States — Indian
J. Phys. $\underline{30}$, 106 (1956) — S.B. Banerjee

UV Absorption in Vapour — Opt. Spektrosk.
$\underline{21}$, 26 (1966) — O.P. Kharitonova

Phosphorescence — J. Mol. Spectrosc. $\underline{36}$,
155 (1970) — J.W. Rabalais, A. Wahlberg
and S.P. McGlynn

UV Absorption in Vapour — Indian J. Pure
Appl. Phys. $\underline{11}$, 37 (1973) — D. Marjit
and S.B. Banerjee

Fluorescence — J. Chem. Phys. $\underline{59}$, 3433
(1973) — A. Reiser and T.R. Wright

β-Methylnaphthalene $(C_{10}H_7CH_3)$

UV Absorption in Vapour — Z. Phys. Chem.
$\underline{11B}$, 369 (1925) — H.G. De'Laszlo

Raman — Z. Phys. Chem. $\underline{78}$, 123 (1932) —
S. Zeimecki

Raman — Z. Phys. Chem. $\underline{202}$, 390 (1954) —
H. Luther and B. Hempel

UV Absorption in Vapour — J. Chem. Phys.
$\underline{23}$, 127 (1955) — H.M. McConnell and
D.D. Tunnicliff

UV Absorption in Solution — J. Chem. Phys.
$\underline{23}$, 927 (1955) — H.M. McConnell and
D.D. Tunnicliff

UV Absorption in Different States — Indian
J. Phys. $\underline{30}$, 106 (1956) — S.B. Banerjee

UV Absorption in Vapour — Indian J. Pure
Appl. Phys. $\underline{4}$, 89 (1966) — R.D. Singh
and R.S. Singh

UV Emission in Vapour — Indian J. Pure
Appl.Phys. $\underline{4}$, 89 (1966) — R.D. Singh
and R.S. Singh

UV Absorption in Vapour — Opt. Spektrosk.
$\underline{21}$, 26 (1966) — O.P. Kharitonova

IR Spectrum — Indian J. Pure Appl. Phys.
$\underline{8}$, 348 (1970) — R.D. Singh and R.S.
Singh

Phosphorescence — J. Mol. Spectrosc. $\underline{36}$,
155 (1970) — J.W. Rabalain, A. Wahlberg
and S.P. McGlynn

UV Absorption in Vapour — Indian J. Pure
Appl.Phys. $\underline{11}$, 37 (1973) — D. Marjit
and S.B. Banerjee

Fluorescence — J. Chem. Phys. $\underline{59}$, 3433
(1973) — A. Reiser and T.R. Wright

2-Methyl-1,4-Naphthoquinone
$(C_{10}H_5O_2CH_3)$

UV Absorption in Solution — J. Chem. Soc.
325 (1935) — K. Macbeth, J.R. Price
and F.L. Winzer

UV Absorption in Solution — J. Chem. Soc.
159 (1941) — R.A. Morton and W.T.
Earlam

UV Absorption in Vapour — Indian J. Pure
Appl. Phys. $\underline{5}$, 430 (1967) — S.Nath
Singh and R.S. Singh

2-Methyl-4-Nitroaniline
($C_6H_3NO_2CH_3NH_2$)

IR Spectrum — Aust. J. Chem. 23, 947
(1970) — L.K. Dyall

o-Methylphenetole ($C_6H_4CH_3OC_2H_5$)

UV Absorption in Different States — Indian
J. Phys. 34, 289 (1960) — K.V.K. Rao
and V.R. Rao
Raman — J. Sci. Ind. Res. (India) 20B, 182
(1961) — K.V.K. Rao and V.R. Rao
UV Absorption in Vapour — Indian J. Pure
Appl. Phys. 1, 20 (1963) — K.V.K. Rao
and V.R. Rao
IR Spectrum — Indian J. Pure Appl. Phys.
1, 51 (1963) — K.V.K. Rao

m-Methylphenetole ($C_6H_4CH_3OC_2H_5$)

UV Absorption in Different States — Indian
J. Phys. 34, 289 (1960) — K.V.K. Rao
and V.R. Rao
Raman — J. Sci. Ind. Res. (India) 20B, 182
(1961) — K.V.K. Rao and V.R. Rao
UV Absorption in Vapour — Indian J. Pure
Appl. Phys. 1, 20 (1963) — K.V.K. Rao
and V.R. Rao
IR Spectrum — Indian J. Pure Appl. Phys.
1, 51 (1963) — K.V.K. Rao

p-Methylphenetole ($C_6H_4CH_3OC_2H_5$)

UV Absorption in Different States — Indian
J. Phys. 34, 289 (1960) — K.V.K. Rao
and V.R. Rao
Raman — J. Sci. Ind. Res. (India) 20B, 182
(1961) — K.V.K. Rao and V.R. Rao
UV Absorption in Vapour — Indian J. Pure
Appl. Phys. 1, 20 (1963) — K.V.K. Rao
and V.R. Rao
IR Spectrum — Indian J. Pure Appl. Phys.
1, 51 (1963) — K.V.K. Rao

Methylphenol ($C_6H_4CH_3OH$)

See   Cresol

2-Methylpyridine ($C_5H_4NCH_3$)

Raman — Indian J. Phys. 10, 23 (1936) —
K. Jatkar

Raman — Z. Phys. Chem. 53B, 124 (1943) —
E. Herz, L. Kahovec and K.W.F. Kohl-
rausch
UV Absorptioon in Vapour — Disc. Faraday
Soc. 9, 26 (1950) — E.F. Herrington
UV Absorption in Vapour — J. Chem. Phys.
20, 1847 (1952) — J.H. Rush and H.
Sponer
UV Absorption in Solution — Trans. Faraday
Soc. 50, 918 (1954) — R.J.L. Andon,
J.D. Cox and E.F. Herrington
UV Absorption in Different States — Indian
J. Phys. 31, 11 (1957) — S.B. Banerjee
Raman — Indian J. Phys. 31, 395 (1957) —
G.S. Kastha
Vibrational Analysis — Spectrochim. Acta
19, 549 (1963) — J.H.S. Green, W.
Kynaston and H.M. Paisley
IR Spectrum — Indian J. Pure Appl. Phys. 9,
346 (1971) — B.R. Pandey and R.S.
Tripathi

3-Methylpyridine ($C_5H_4NCH_3$)

Raman — Indian J. Phys. 10, 23 (1936) —
K. Jatkar
Raman — Z. Phys. Chem. 53B, 124 (1943) —
E. Herz, L. Kahovec and K.W.F.
Kohlrausch
UV Absorption in Vapour — Disc. Faraday
Soc. 9, 26 (1950) — E.F. Herrington
UV Absorption in Vapour — J. Chem. Phys.
20, 1847 (1952) — J.H. Rush and H.
Sponer
UV Absorption in Solution — Trans. Faraday
Soc. 50, 918 (1954) — R.J.L. Andon,
J.D. Cox and E.F. Herrington
UV Absorption in Different States — Indian
J. Phys. 31, 11 (1957) — S.B. Banerjee
Raman — Indian J. Phys. 31, 395 (1957) —
G.S. Kastha
Vibrational Analysis — Spectrochim. Acta
19, 549 (1963) — J.H.S. Green, W.
Kynaston and H.M. Paisley
IR Spectrum — Indian J. Pure Appl. Phys.
9, 346 (1971) — B.R. Pandey and R.S.
Tripathi

4-Methylpyridine ($C_5H_4NCH_3$)

Raman — Z. Phys. Chem. 53B, 124 (1943) —
E. Herz, L. Kahovec and K.W.F.
Kohlrausch
UV Absorption in Vapour — Disc. Faraday
Soc. 9, 26 (1950) — E.F. Herrington
UV Absorption in Vapour — J. Chem. Phys.
20, 1847 (1952) — J.H.Rush and H.
Sponer

UV Absorption in Solution — Trans. Faraday
    Soc. 50, 918 (1954) — R.J.L. Andon,
    J.D. Cox and E.F. Herrington
Raman — Indian J. Phys. 31, 395 (1957) —
    G.S. Kastha
UV Absorption in Different States — Indian
    J. Phys. 31, 11 (1957) — S.B. Banerjee
Vibrational Analysis — Trans. Faraday Soc.
    53, 1171 (1957) — D.A. Long, F.S.
    Musfin, J.L. Hales and W. Kynaston
Vibrational Analysis — Spectrochim. Acta
    19, 549 (1963) — J.H.S. Green, W.
    Kynaston and H.M. Paisley
Vibrational Analysis — Spectrochim. Acta
    19, 1777 (1963) — D.A. Long and W.O.
    George
IR Spectrum — Indian J. Pure Appl. Phys.
    9, 346 (1971) — B.R. Pandey and R.S.
    Tripathi
Calculation — J. Mol. Struct. 32, 67
    (1976) — J.B. Moffat

4-Methylpyridine-D₇ (C₅D₄NCD₃)

Vibrational Analysis — Spectrochim. Acta
    19, 1777 (1963) — D.A. Long and W.O.
    George

4-Methylpyrimidine (C₄H₃N₂CH₃)

UV Absorption in Vapour — Indian J. Pure
    Appl. Phys. 12, 726 (1974) — M.A.
    Shashidhar

5-Methylpyrimidine (C₄H₃N₂CH₃)

IR Spectrum — Spectrochim. Acta 9, 113
    (1957) — R.C. Lord, A.L. Marston and
    F.A. Miller
UV Absorption in Vapour — Curr. Sci. 42,
    816 (1973) — M.A. Shashidhar and K.S.
    Rao
UV Absorption in Vapour — Indian J. Phys.
    49, 128 (1975) — M.A. Shashidhar and
    K.S. Rao
IR Spectrum — Indian J. Phys. 49, 128
    (1975) — M.A. Shashidhar and K.S. Rao

2-Methylquinoline (C₉H₆NCH₃)

Raman — Indian J. Phys. 10, 23 (1936) —
    K. Jatkar
UV Absorption in Solution — J. Amer. Chem.
    Soc. 69, 14 (1947) — P.L. Pikard and
    H.L. Lochte

Raman — Z. Phys. Chem. 195, 103 (1950) —
    H. Luther and C. Reichel
UV Absorption in Solution — J. Amer. Chem.
    Soc. 77, 2577 (1955) — S.B. Knight,
    R.H. Wallick and R. Bloch
IR Spectrum — Chem. Pharm. Bull.(Tokyo)
    4, 292 (1956) — H. Sindo and N.
    Ikeawa
IR Spectrum — Chem. Pharm. Bull.(Tokyo)
    8, 856 (1960) — H. Sindo
IR Spectrum — J. Chem. Soc. 2942 (1960) —
    A.R. Katritzky and R.A. Jones
UV Absorption in Vapour — Curr. Sci. 35,
    230 (1966) — M.A. Shashidhar and
    K.S. Rao
UV Absorption in Vapour — Indian J. Phys.
    41, 299 (1967) — M.A. Shashidhar and
    K.S. Rao

4-Methylquinoline (C₉H₆NCH₃)

Raman — Indian J. Phys. 10, 23 (1936) —
    K. Jatkar
UV Absorption in Solution — J. Amer. Chem.
    Soc. 69, 14 (1947) — P.L. Pikard
    and H.L. Lochte
Raman — Z. Phys. Chem. 195, 103 (1950) —
    H. Luther and C. Reichel
UV Absorption in Solution — J. Amer. Chem.
    Soc. 77, 2577 (1955) — S.B. Knight,
    R.H. Wallick and R. Bloch
IR Spectrum — Chem. Pharm. Bull. (Tokyo)
    4, 292 (1956) — H. Sindo and N. Ikeawa
IR Spectrum — Chem. Pharm. Bull. (Tokyo)
    8, 845 (1960) — H. Sindo
IR Spectrum — J. Chem. Soc. 2942 (1960) —
    A.R. Katritzky and R.A. Jones
UV Absorption in Vapour — Curr. Sci. 35,
    230 (1966) — M.A. Shashidhar and K.S.
    Rao
UV Absorption in Vapour — Indian J. Phys.
    41, 299 (1967) — M.A. Shashidhar
    and K.S. Rao

6-Methylquinoline (C₉H₆NCH₃)

UV Absorption in Solution — J. Amer. Chem.
    Soc. 69, 14 (1947) — P.L. Pikard and
    H.L. Lochte
UV Absorption in Solution — J. Amer. Chem.
    Soc. 77, 2577 (1955) — S.B. Knight,
    R.H. Wallick and R. Bloch
IR Spectrum — Chem. Pharm. Bull. (Tokyo)
    4, 292 (1956) — H. Sindo and N. Ikeawa
IR Spectrum — Chem. Pharm. Bull. (Tokyo)
    8, 845 (1960) — H. Sindo

IR Spectrum — J. Chem. Soc. 2942 (1960) —
A.R. Katritzky and R.A. Jones

UV Absorption in Vapour — Curr. Sci. 35,
230 (1966) — M.A. Shashidhar and K.S.
Rao

UV Absorption in Vapour — Indian J. Phys.
41, 299 (1967) — M.A. Shashidhar and
K.S. Rao

7-Methylquinoline ($C_9H_6NCH_3$)

UV Absorption in Solution — J. Amer. Chem.
Soc. 69, 14 (1947) — P.L. Pikard and
H.L. Lochte

UV Absorption in Solution — J. Amer. Chem.
Soc. 77, 2577 (1955) — S.B. Knight,
R.H. Wallick and R. Bloch

IR Spectrum — Chem. Pharm. Bull. (Tokyo)
4, 292 (1956) — H. Sindo and N. Ikeawa

IR Spectrum — Chem. Pharm. Bull. (Tokyo)
8, 845 (1960) — H. Sindo

IR Spectrum — J. Chem. Soc. 2942 (1960) —
A.R. Katritzky and R.A. Jones

UV Absorption in Vapour — Curr. Sci. 35, 230
(1966) — M.A. Shashidhar and K.S. Rao

UV Absorption in Vapour — Indian J. Phys.
41, 299 (1967) — M.A. Shashidhar and
K.S. Rao

o-Methylstyrene ($C_6H_4CH_3CH:CH_2$)

UV Absorption in Solution — Bull. Soc. Chim.
Fr. 5, 848 (1938) — P. Ramart-Lucas
and J. Hock

UV Absorption in Solution — J. Amer. Chem.
Soc. 71, 3241 (1943) — Y. Hirschberg

IR Spectrum — J. Appl. Phys. 16, 77 (1945) —
R.B. Barnes, U. Liddel and V.Z. Williams

UV Absorption in Solution — J. Amer. Chem.
Soc. 69, 2707 (1947) — H.A. Laitinen,
F.A. Miller and T.D. Parks

IR Spectrum — J. Chem. Soc. 2389 (1949) —
K.C. Bryant, G.T. Kennedy and E.M.
Tanner

Raman — Zh. Fiz. Khim. 36, 356 (1962) —
Yu-Pao-Shan, V.N. Nikitin and M.V. Vol'-
kenshtein

UV Absorption in Vapour — Indian J. Pure
Appl. Phys. 6, 614 (1968) — B.J. Ansari
and D. Sharma

Vibrational Analysis — Appl. Spectrosc. 22,
650 (1968) — W.G. Fateley, G.L. Carlson
and F.E. Dickson

IR Spectrum — Indian J. Phys. 44, 25 (1970) —
V.N. Verma and K. Singh

IR Spectrum — Indian J. Pure Appl. Phys.
8, 725 (1970) — B.J. Ansari

m-Methylstyrene ($C_6H_4CH_3CH:CH_2$)

UV Absorption in Solution — Bull. Soc.
Chim. Fr. 5, 848 (1938) — P. Ramart-
Lucas and J. Hock

UV Absorption in Solution — J. Amer. Chem.
Soc. 71, 3241 (1943) — Y. Hirschberg

IR Spectrum — J. Appl. Phys. 16, 77
(1945) — R.B. Barnes, U. Liddel and
V.Z. Williams

UV Absorption in Solution — J. Amer. Chem.
Soc. 69, 2707 (1947) — H.A. Laitinen,
F.A. Miller and T.D. Parks

IR Spectrum — J. Chem. Soc. 2389 (1949) —
K.C. Bryant, G.T. Kennedy and E.M.
Tanner

Raman — Zh. Fiz. Khim. 36, 356 (1962) —
Yu-Pao-Shan, V.N. Nikitin and M.V.
Vol'kenshtein

UV Absorption in Vapour — Indian J. Pure
Appl. Phys. 6, 614 (1968) — B.J.
Ansari and D. Sharma

Vibrational Analysis — Appl. Spectrosc. 22,
650 (1968) — W.G. Fateley, G.L. Carlson
and F.E. Dickson

IR Spectrum — Indian J. Phys. 44, 25
(1970) — V.N. Verma and K. Singh

IR Spectrum — Indian J. Pure Appl. Phys.
8, 725 (1970) — B.J. Ansari

p-Methylstyrene ($C_6H_4CH_3CH:CH_2$)

UV Absorption in Solution — Bull. Soc.
Chim. Fr. 5, 848 (1938) — P. Ramart-
Lucas and J. Hock

UV Absorption in Solution — J. Amer. Chem.
Soc. 71, 3241 (1943) — Y. Hirschberg

IR Spectrum — J. Appl. Phys. 16, 77
(1945) — R.B. Barnes, U. Liddel and
V.Z. Williams

UV Absorption in Solution — J. Amer. Chem.
Soc. 69, 2707 (1947) — H.A. Laitinen,
F.A. Miller and T.D. Parks

IR Spectrum — J. Chem. Soc. 2389 (1949) —
K.C. Bryant, G.T. Kennedy and E.M.
Tanner

Raman — Zh. Fiz. Khim. 36, 356 (1962) —
Yu-Pao-Shan, V.N. Nikitin and M.V.
Vol'kenshtein

UV Absorption in Vapour — Indian J. Pure
Appl. Phys. 6, 614 (1968) — B.J.
Ansari and D. Sharma

IR Spectrum — Indian J. Pure Appl. Phys.
8, 725 (1970) — B.J. Ansari

Raman — J. Mol. Struct. 37, 85 (1977) —
L.A. Carreira and T.G. Towns

α-Naphthaldehyde ($C_{10}H_7CHO$)

IR Spectrum – J. Amer. Chem. Soc. 72, 5226 (1950) – M. Hunsberger

IR Spectrum – J. Amer. Chem. Soc. 72, 526 (1950) – H. Moyer

IR Spectrum – Anal. Chem. 29, 334 (1957) – P. Sargas and S. Pinchas

IR Spectrum – Spectrochim. Acta 15, 1118 (1959) – T.S. Wang and J.M. Sanders

Phosphorescence – J. Amer. Chem. Soc. 88, 50 (1966) – D.R. Kearns ans W.A. Case

Vibrational Analysis – Indian J. Pure Appl. Phys. 7, 649 (1969) – K. Singh, S. Nath Singh and D.K. Rai

Vibrational Analysis – Indian J. Phys. 48, 494 (1974) – O.P. Sharma, S.N. Singh and R.D. Singh

Calculation – J. Mol. Spectrosc. 61, 350 (1976) – S. Suzuki and T. Fujii

β-Naphthaldehyde ($C_{10}H_7CHO$)

IR Spectrum – J. Amer. Chem. Soc. 72, 5226 (1950) – M. Hunsberger

IR Spectrum – J. Amer. Chem. Soc. 72, 526 (1950) – H. Moyer

IR Spectrum – Anal. Chem. 29, 334 (1957) – P. Sargas and S. Pinchas

IR Spectrum – Spectrochim. Acta 15, 1118 (1959) – T.S. Wang and J.M. Sanders

Phosphorescence – J. Amer. Chem. Soc. 88, 50 (1966) – D.R. Kearns and W.A. Case

Vibrational Analysis – Indian J. Pure Appl. Phys. 7, 649 (1969) – K. Singh, S. Nath Singh and D.K. Rai

Vibrational Analysis – Indian J. Phys. 48, 494 (1974) – O.P. Sharma, S.N. Singh and R.D. Singh

Calculation – J. Mol. Spectrosc. 61, 350 (1976) – S. Suzuki and T. Fujii

Naphthalene ($C_{10}H_8$)

UV Absorption in Vapour – Proc. Roy. Soc. A105, 662 (1924) – V. Henri and H.G. De Laszlo

Vibrational Analysis – Proc. Indian Acad. Sci A15, 250 (1942) – R. Norris

Vibrational Analysis – Proc. Indian Acad. Sci. A15, 376 (1942) – T.M.K. Nedumgadi

Phosphorescence – J. Amer. Chem. Soc. 66, 2100 (1944) – G.N. Lears and M. Kasha

Vibrational Analysis – Z. Elektrochem. 52, 210 (1948) – H. Luther

Vibrational Analysis – Anal. Chem. 22, 1074 (1950) – W.G. Braun, D.F. Spooner and M.R. Fanke

UV Absorption in Vapour – Disc. Faraday Soc. 9, 19 (1950) – H. Sponer and P. Nordheim

Raman (Crystal) – Indian J. Phys. 24, 539 (1950) – A.K. Ray

Fluorescence – J. Chem. Phys. 20, 1375 (1952) – O.Schnepp and D.S. McClure

Vibrational Analysis – J. Chem. Phys. 20, 270 (1952) – G.C. Pimentel and A.L. McClellan

Vibrational Analysis – Z. Phys. Chem. 202, 390 (1954) – H. Luther and B. Hampel

Vibrational Analysis – Z. Elektrochem. 59, 1012 (1955) – H. Luther, G. Brandes, H. Gunzler and B. Hampel

Vibrational Analysis – J. Chem. Phys. 23, 238 (1955) – E.R. Lippincott and E.J. O'Reily

IR Spectrum – J. Chem. Phys. 23, 230 (1955) – W.B. Pearson, G.C. Pimentel and O. Schnepp

Vibrational Analysis – Z. Elektrochem. 59, 1008 (1955) – H. Luther, K. Feldman and B. Hampel

Vibrational Analysis – J. Chem. Phys. 23, 245 (1955) – A.L. McClellan and G.C. Pimentel

UV Absorption in Vapour – J. Chem. Phys. 23, 646 (1955) – H. Sponer and C.D. Cooper

Vibrational Analysis – Z. Phys. 144, 428 (1956) – J. Brandmuller and E.W. Schmid

Vibrational Analysis – J. Mol. Spectrosc. 1, 257 (1957) – D.B. Scully and D.H. Whiffen

Calculation – Z. Elektrochem. 62, 1005 (1958) – E.W. Schmid

IR Spectrum – Can. J. Phys. 37, 553 (1959) – S.S. Mitra and H.J. Bernstein

Fluorescence – Opt. Spektrosk. 7, 24 (1959) – T.N. Bolotnukova

Calculation – Spectrochim. Acta 16, 1393 (1960) – D.E. Freeman and I.G. Ross

UV Absorption in Vapour – J. Chem. Phys. 33, 892 (1960) – O.E. Weigang

IR Spectrum – Spectrochim. Acta 16, 58 (1960) – E.R. Lippincott, C.E. Weir, A. van Valkenburg and E.N. Bunting

Calculation – Spectrochim. Acta 16, 1409 (1960) – D.B. Scully and D.H. Whiffen

IR Spectrum – Spectrochim. Acta 16, 1393 (1960) – D.E. Freeman and I.G. Ross

IR Spectrum – Spectrochim. Acta 16, 74 (1960) – A.G. Moritz

UV Absorption Spectrum – J. Mol. Spectrosc. 6, 305 (1961) – D.E. Freeman

UV Absorption in Vapour – Philos. Trans. R. Soc. Lond. A253, 1543 (1961) – D.P. Craig, J.M. Hollas, M.F. Redies and S.C. Wait, Jr

UV Emission in Vapour – J. Mol. Spectrosc. 6, 305 (1961) – D.E. Freeman

Fluorescence – J. Mol. Spectrosc. 9, 138 (1962) – J.M. Hollas

Fluorescence – Spectrochim. Acta 18, 425 (1962) – E. Hutton and B. Stevens

Vibrational Analysis – Z. Elektrochem. 66, 546 (1962) – H. Luther and H.J. Drewitz

Vibrational Analysis – J. Chem. Phys. 36, 3308 (1962) – J.R. Scherer

Fluorescence – J. Mol. Spectrosc. 9, 138 (1962) – J.M. Hollas

IR Spectrum – J. Mol. Spectrosc. 10, 79 (1963) – S.C. Wait, Jr and F.E. Shaffer

Calculation – Spectrochim. Acta 20, 891 (1964) – D.J. Evans and D.B. Scully

UV Absorption in Vapour (High Resolution) – J. Mol. Spectrosc. 16, 406 (1965) – K.K. Innes, J.E. Parkin, D.K. Ervin, J.M. Hollas and I.G. Ross

Calculation – Spectrochim. Acta 22, 1981 (1966) – N. Neto, M. Scrocco and S. Califano

Fluorescence – J. Chem. Phys. 44, 2423 (1966) – R.J. Watts and S.J. Strickler

Phosphorescence – J. Chem. Phys. 49, 3184 (1968) – S.C. Tsai and G.W. Robinson

Absorption Spectrum in Vapour – J. Mol. Spectrosc. 26, 67 (1968) – G.A. George and G.C. Morris

Fluorescence – J. Mol. Spectrosc. 26, 117 (1968) – J.E. Haebig

Phosphorescence – J. Chem. Phys. 51, 5063 (1969) – J.M. Hanson

Vibrational Analysis – Spectrochim. Acta 26A, 1791 (1970) – A.V. Bree and R.A. Kidd

Phosphorescence – J. Mol. Spectrosc. 36, 155 (1970) – J.W. Robalais, A. Wahlborg and S.P. McGlynn

Fluorescence – Chem. Phys. Lett. 7, 306 (1970) – J.O. Uy and E.C. Lim

Fluorescence – Chem. Phys. Lett. 8, 436 (1971) – J.M. Brondeau and M. Stockburger

Fluorescence – J. Mol. Spectrosc. 37, 571 (1971) – S.K. Chakrabarti

Fluorescence – J. Chem. Phys. 54, 1054 (1971) – U. Laor and P.K. Ludwig

Fluorescence – Chem. Phys. Lett. 10, 193 (1971) – P. Wannier, P.M. Rentzepis and J. Jortner

Absorption Spectrum – Aust. J. Chem. 24, 1107 (1971) – J.P. Byrne and I.G. Ross

Fluorescence – Chem. Phys. Lett. 11, 474 (1971) – E.W. Schlag, S. Schneider and D.W. Chandler

Phosphorescence – J. Chem. Phys. 54, 1 (1971) – J. Langlaar, R.P.H. Reltschnick and G.J. Hoytink

Phosphorescence – Chem. Phys. Lett. 13, 662 (1972) – W.H. Van Leeuwen, J. Langlaar and J.D. Van Voorst

Fluorescence – J. Chem. Phys. 56, 3374 (1972) – E.C. Lim and J.O. Uy

Fluorescence – Chem. Phys. Lett. 18, 481 (1973) – G.S. Beddard, G.R. Fleming, O.L.J. Gijzeman and G. Porter

Fluorescence – Chem. Phys. Lett. 19, 29 (1973) – T. Dienum, C.J. Werkhoven, J. Langlaar, R.P.H. Rettschnick and J.D.W. Van Voorst

Fluorescence – J. Chem. Phys. 58, 4686 (1973) – F. Hirayama, T.A. Gregary and S. Lipsky

Fluorescence – Chem. Phys. Lett. 22, 235 (1973) – G.S. Beddard, J.J. Formosinho and G. Porter

Fluorescence – Adv. Chem. Phys. 23, 189 (1973) – P.M. Rentzepis

Fluorescence – J. Chem. Phys. 59, 3433 (1973) – A. Reiser and T.R. Wright

Fluorescence – Proc. Roy. Soc. A340, 519 (1974) – G.S. Beddard, G.R. Fleming, O.L.J. Gijzeman and G. Porter

Calculation – J. Mol. Spectrosc. 50, 457 (1974) – J.E. Ridley and M.C. Zerner

Excitation Spectrum (Two Photon) – Chem. Phys. Lett. 26, 323 (1974) – A. Bergman and J. Jortner

Excitation Spectrum (Two Photon) – Chem. Phys. Lett. 31, 472 (1975) – N. Mikami and M. Ito

Fluorescence – Chem. Phys. Lett. 31, 1 (1975) – U. Boesl, H.J. Neasser and E.W. Schlag

Fluorescence (Electron Impact) – J. Chem. Phys. 62, 136 (1975) – K.C. Smith, J.A. Schiavone and R.S. Freund

Naphthalene-D_8 ($C_{10}D_8$)

Vibrational Analysis – J. Chem. Phys. 23, 230 (1955) – W.B. Pearson, G.C. Pimentel and O. Schnepp

Vibrational Analysis – J. Chem. Phys. 23, 238 (1955) – E.R. Lippincott and E.J. O'Reilly

Vibrational Analysis – Spectrochim. Acta 10, 1465 (1955) – G.W. Chantry, A. Anderson and H.A. Gebbie

Vibrational Analysis – J. Chem. Phys. 23, 245 (1955) – A.L. McClellan and G.C. Pimentel

Vibrational Analysis – Z. Elektrochem. <u>59</u>, 1012 (1955) – H. Luther, G. Brandes, H. Gunzler and B. Hampel

Vibrational Analysis – Spectrochim. Acta <u>16</u>, 1393 (1960) – D.E. Freeman and I.G. Ross

Calculation – J. Chem. Phys. <u>36</u>, 3308 (1962) – J.R. Scherer

Fluorescence – J. Mol. Spectrosc. <u>9</u>, 138 (1962) – J.M. Hollas

Calculation – Spectrochim. Acta <u>22</u>, 1981 (1966) – N. Neto, M. Scrocco and S. Califano

Phosphorescence – J. Chem. Phys. <u>49</u>, 3184 (1968) – S.C. Tsai and G.W. Robinson

Vibrational Analysis – Spectrochim. Acta <u>26A</u>, 1791 (1970) – A.V. Bree and R.A. Kydd

Excitation Spectrum (Two Photon) – Chem. Phys. Lett. <u>31</u>, 472 (1975) – N. Mikami and M. Ito

Fluorescence – Chem. Phys. Lett. <u>31</u>, 7 (1975) – U. Boesl, H.J. Neasser and E.W. Schlag

1,3-Naphthalenediol ($C_{10}H_6(OH)_2$)

Absorption Spectrum – Spectrochim. Acta <u>15</u>, 393 (1959) – D.M. Hercules and L.B. Rogers

Fluorescence – Spectrochim. Acta <u>15</u>, 393 (1959) – D.M. Hercules and L.B. Rogers

1,4-Naphthalenediol ($C_{10}H_6(OH)_2$)

Absorption Spectrum – Spectrochim. Acta <u>15</u>, 393 (1959) – D.M. Hercules and L.B. Rogers

Fluorescence – Spectrochim. Acta <u>15</u>, 393 (1959) – D.M. Hercules and L.B. Rogers

1,5-Naphthalenediol ($C_{10}H_6(OH)_2$)

Absorption Spectrum – Spectrochim. Acta <u>15</u>, 393 (1959) – D.M. Hercules and L.B. Rogers

Fluorescence – Spectrochim. Acta <u>15</u>, 393 (1959) – D.M. Hercules and L.B. Rogers

Absorption Spectrum – J. Mol. Spectrosc. <u>50</u>, 90 (1974) – K. Hara, T. Takemura and H. Baba

1,6-Naphthalenediol ($C_{10}H_6(OH)_2$)

Absorption Spectrum – Spectrochim. Acta <u>15</u>, 393 (1959) – D.M. Hercules and L.B. Rogers

Fluorescence – Spectrochim. Acta <u>15</u>, 393 (1959) – D.M. Hercules and L.B. Rogers

2,3-Naphthalenediol ($C_{10}H_6(OH)_2$)

Absorption Spectrum – Spectrochim. Acta <u>15</u>, 393 (1959) – D.M. Hercules and L.B. Rogers

Fluorescence – Spectrochim. Acta <u>15</u>, 393 (1959) – D.M. Hercules and L.B. Rogers

Absorption Spectrum – J. Mol. Spectrosc. <u>50</u>, 90 (1974) – K. Hara, T. Takemura and H. Baba

2,6-Naphthalenediol ($C_{10}H_6(OH)_2$)

Absorption Spectrum – Spectrochim. Acta <u>15</u>, 393 (1959) – D.M. Hercules and L.B. Rogers

Fluorescence – Spectrochim. Acta <u>15</u>, 393 (1959) – D.M. Hercules and L.B. Rogers

2,7-Naphthalenediol ($C_{10}H_6(OH)_2$)

Absorption Spectrum – Spectrochim. Acta <u>15</u>, 393 (1959) – D.M. Hercules and L.B. Rogers

Fluorescence – Spectrochim. Acta <u>15</u>, 393 (1959) – D.M. Hercules and L.B. Rogers

Absorption Spectrum – J. Mol. Spectrosc. <u>50</u>, 90 (1974) – K. Hara, T. Takemura and H. Baba

Naphthazarin ($C_{10}H_2(OH)_4$)

Vibrational Analysis – Trans. Faraday Soc. <u>50</u>, 911 (1954) – D. Hadzi and N. Sheppard

UV Absorption in Vapour – Indian J. Pure Appl. Phys. <u>7</u>, 283 (1969) – G.D. Baruah, S. Nath Singh and R.S. Singh

Vibrational Analysis – Indian J. Pure Appl. Phys. <u>7</u>, 283 (1969) – G.D. Baruah, S. Nath Singh and R.S. Singh

α-Naphthol ($C_{10}H_7OH$)

Absorption Spectrum – Spectrochim. Acta <u>15</u>, 393 (1959) – D.M. Hercules and L.B. Rogers

Fluorescence — Spectrochim. Acta $\underline{15}$, 393
    (1959) — D.M. Hercules and L.B. Rogers
UV Absorption in Vapour — Indian J. Pure
    Appl. Phys. $\underline{3}$, 418 (1965) — R.D. Singh
    and R.S. Singh
IR Spectrum — Indian J. Pure Appl. Phys.
    $\underline{8}$, 348 (1970) — R.D. Singh and R.S.
    Singh
Phosphorescence — J. Mol. Struct. $\underline{7}$, 455
    (1971) — S.K. Chakrabarty
Fluorescence — J. Mol. Struct. $\underline{7}$, 455
    (1971) — S.K. Chakrabarty
Absorption Spectrum — J. Mol. Spectrosc.
    $\underline{47}$, 243 (1973) — S. Suzuki, T. Fujii
    and H. Baba
Calculation — J. Mol. Spectrosc. $\underline{61}$, 350
    (1976) — S. Suzuki and T. Fujii
Vibrational Analysis — Indian J. Phys. $\underline{51B}$,
    93 (1977) — O.P. Sharma and R.D. Singh

β-Naphthol ($C_{10}H_7OH$)

Absorption Spectrum — Spectrochim. Acta
    $\underline{15}$, 393 (1959) — D.M. Hercules and L.B.
    Rogers
Fluorescence — Spectrochim. Acta $\underline{15}$, 393
    (1959) — D.M. Hercules and L.B. Rogers
UV Absorption in Vapour — Indian J. Pure
    Appl. Phys. $\underline{3}$, 418 (1965) — R.D. Singh
    and R.S. Singh
UV Emission in Vapour — Curr. Sci. $\underline{35}$, 386
    (1966) — R.D. Singh and R.S. Singh
IR Spectrum — Indian J. Pure Appl. Phys.
    $\underline{8}$, 348 (1970) — R.D. Singh and R.S.
    Singh
Phosphorescence — J. Mol. Struct. $\underline{7}$, 455
    (1971) — S.K. Chakrabarty
Fluorescence — J. Mol. Struct. $\underline{7}$, 455
    (1971) — S.K. Chakrabarty
Absorption Spectrum — J. Mol. Spectrosc.
    $\underline{47}$, 243 (1973) — S. Suzuki, T. Fujii
    and H. Baba
Fluorescence — Opt. Spektrosk. $\underline{10}$, 283
    (1974) — L. Lindquist, R. Lopez-Degado,
    M.M. Martin and A. Tramer
Calculation — J. Mol. Spectrosc. $\underline{61}$, 350
    (1976) — S. Suzuki and T. Fujii
Vibrational Analysis — Indian J. Phys. $\underline{51B}$,
    93 (1977) — O.P. Sharma and R.D. Singh

1,4-Naphthoquinone ($C_{10}H_6O_2$)

UV Absorption in Vapour — J. Chem. Soc.
    1841 (1923) — J.E. Purvis
UV Absorption in Vapour — Indian Acad. Sci.
    $\underline{3A}$, 173 (1936) — P.K. Seshan

UV Absorption in Solution — Z. Naturforsch.
    $\underline{79}$, 360 (1952) — H. Hartmann and E.
    Lorenz
IR Spectrum — Trans. Faraday Soc. $\underline{50}$, 911
    (1954) — D. Hadzi and N. Sheppard
UV Absorption in Solution — Z. Naturforsch.
    $\underline{10}$, 668 (1955) — W. Fraig, Th. Ploretz
    and A.K. Kullmer
UV Emission in Vapour — Indian J. Pure
    Appl. Phys. $\underline{5}$, 394 (1967) — S. Nath
    Singh and R.S. Singh
UV Absorption in Vapour — Indian J. Pure
    Appl. Phys. $\underline{6}$, 187 (1968) — S. Nath
    Singh and R.S. Singh
UV Emission in Vapour — Indian J. Pure
    Appl. Phys. $\underline{6}$, 91 (1968) — R.S. Singh
UV Emission in Vapour — Indian J. Pure
    Appl. Phys. $\underline{11}$, 303 (1973) — S. Nath
    Singh
Visible Absorption in Vapour — Indian J.
    Pure Appl. Phys. $\underline{11}$, 303 (1973) —
    S. Nath Singh

α-Naphthylamine ($C_{10}H_7NH_2$)

IR Spectrum — Spectrochim. Acta $\underline{15}$, 1118
    (1959) — T.S. Wang and J.M. Sanders
Calculation — Spectrochim. Acta $\underline{19}$, 705
    (1963) — P.J. Krueger
IR Spectrum — J. Chim. Phys. $\underline{66}$, 889
    (1964) — C. Nicole and C. Garrigou-
    Lagrange
Phosphorescence — J. Mol. Struct. $\underline{7}$, 455
    (1971) — S.K. Chakrabarty
Fluorescence — J. Mol. Struct. $\underline{7}$, 455
    (1971) — S.K. Chakrabarty
Absorption Spectrum — J. Mol. Spectrosc.
    $\underline{47}$, 243 (1973) — S. Suzuki, T. Fujii
    and H. Baba
Calculation — J. Mol. Spectrosc. $\underline{61}$, 350
    (1976) — S. Suzuki and T. Fujii

β-Naphthylamine ($C_{10}H_7NH_2$)

IR Spectrum — Spectrochim. Acta $\underline{15}$, 1118
    (1959) — T.S. Wang and J.M. Sanders
Calculation — Spectrochim. Acta $\underline{19}$, 705
    (1963) — P.J. Krueger
IR Spectrum — J. Chim. Phys. $\underline{66}$, 889
    (1964) — C. Nicole and C. Garrigou-
    Lagrange
Fluorescence — J. Chem. Phys. $\underline{51}$, 2508
    (1969) — E.W. Schlag and H.V. Weyssen-
    hoff
Phosphorescence — J. Mol. Struct. $\underline{7}$, 455
    (1971) — S.K. Chakrabarty

Fluorescence — Chem. Phys. Lett. <u>17</u>, 309
(1972) — Th. Forster
Absorption Spectrum — J. Mol. Spectrosc.
<u>47</u>, 243 (1973) — S. Suzuki, T. Fujii
and H. Baba
Calculation — J. Mol. Spectrosc. <u>61</u>, 350
(1976) — S. Suzuki and T. Fujii

1,5-Naphthyridine $(C_8H_6N_2)$

IR Spectrum — J. Mol. Spectrosc. <u>35</u>, 251
(1970) — J.A. Merritt and R.J. Pirkle

1,6-Naphthyridine $(C_8H_6N_2)$

IR Spectrum — Chem. Pharm. Bull.(Tokyo)
<u>6</u>, 404 (1958) — N. Ikewana
IR Spectrum — Spectrochim. Acta <u>22</u>, 117
(1966) — W.L.F. Armarego, G.B. Barlin
and E. Spinner
Vibrational Analysis — J. Mol. Spectrosc.
<u>46</u>, 401 (1973) — J.T. Carrano and S.C.
Wait, Jr

1,6-Naphthyridine-D_6 $(C_8D_6N_2)$

Absorption Spectrum — J. Mol. Spectrosc.
<u>49</u>, 201 (1974) — G. Fischer

1,8-Naphthyridine $(C_8H_6N_2)$

IR Spectrum — Chem. Pharm. Bull. (Tokyo)
<u>6</u>, 404 (1958) — N. Ikewana
IR Spectrum — Spectrochim. Acta <u>22</u>, 117
(1966) — W.L.F. Armarego, G.B. Barlin
and E. Spinner
Vibrational Analysis — J. Mol. Spectrosc.
<u>46</u>, 401 (1973) — J.T. Carrano and
S.C. Wait, Jr

o-Nitroaniline $(C_6H_4NH_2NO_2)$

Vibrational Analysis — J. Chem. Soc. 2051
(1951) — A.R. Katritzky and P. Simons
Vibrational Analysis — J. Chem. Soc. 4183
(1952) — R.R. Randle and D.H. Whiffen
Vibrational Analysis — J. Amer. Chem. Soc.
<u>74</u>, 1265 (1952) — R.J. Francel
IR Spectrum — Aust. J. Chem. <u>11</u>, 513 (1958)—
L.K. Dyall and A.N. Hambly
UV Absorption in Solution — Can. J. Chem.
<u>38</u>, 1837 (1960) — J.C. Dearden and
W.F. Forbes

IR Spectrum — Spectrochim. Acta <u>17</u>, 291
(1961) — L.K. Dyall
IR Spectrum — Indian J. Phys. <u>37</u>, 275
(1963) — K.C. Medhi and G.S. Kastha
Vibrational Analysis — Indian J. Pure Appl.
Phys. <u>8</u>, 857 (1970) — S.C. Srivastava,
M.P. Srivastava and I.S. Singh
Fluorescence — J. Chem. Phys. <u>58</u>, 1607
(1973) — O.S. Khalil, C.J. Seliskar
and S.P. McGlynn
Raman (Inverse) — Chem. Phys. Lett. <u>29</u>, 389
(1974) — S.H. Lin, E.S. Reid and C.J.
Tredwell
UV Absorption in Vapour — Spectrochim. Acta
<u>33A</u>, 21 (1977) — S. Millefiori, G.
Favini, A. Millefiori and D. Grasso

m-Nitroaniline $(C_6H_4NH_2NO_2)$

Vibrational Analysis — J. Chem. Soc. 2051
(1951) — A.R. Katritzky and P. Simons
Vibrational Analysis — J. Amer. Chem. Soc.
<u>74</u>, 1265 (1952) — R.J. Francel
Vibrational Analysis — J. Chem. Soc. 4183
(1952) — R.R. Randle and D.H. Whiffen
IR Spectrum — J. Amer. Chem. Soc. <u>77</u>, 6341
(1955) — J.F. Brown, Jr
IR Spectrum — Aust. J. Chem. <u>11</u>, 513 (1958)—
L.K. Dyall and A.N. Hambly
UV Absorption in Solution — Can. J. Chem.
<u>38</u>, 1837 (1960) — J.C. Dearden and
W.F. Forbes
IR Spectrum — Indian J. Phys. <u>37</u>, 275
(1963) — K.C. Medhi and G.S. Kastha
Vibrational Analysis — J. Chim. Phys. <u>63</u>,
552 (1966) — C. Garrigou-Lagrange, M.
Chehata and J. Lascombe
IR Spectrum — Spectrochim. Acta <u>25A</u>, 1423
(1969) — L.K. Dyall
Vibrational Analysis — Indian J. Pure Appl.
Phys. <u>8</u>, 857 (1970) — S.C. Srivastava,
M.P. Srivastava and I.S. Singh
Fluorescence — J. Chem. Phys. <u>58</u>, 1607
(1973) — O.S. Khalil, C.J. Seliskar
and S.P. McGlynn
UV Absorption in Vapour — Spectrochim. Acta
<u>33A</u>, 21 (1977) — S. Millefiori, G.
Favini, A. Millefiori and D. Grasso

p-Nitroaniline $(C_6H_4NH_2NO_2)$

Vibrational Analysis — J. Chem. Soc. 2051
(1951) — A.R. Katritzky and P. Simons
Vibrational Analysis — J. Chem. Soc. 4183
(1952) — R.R. Randle and D.H. Whiffen
Vibrational Analysis — J. Amer. Chem. Soc.
<u>74</u>, 1265 (1952) — R.J. Francel

IR Spectrum – J. Amer. Chem. Soc. 77, 6341
(1955) – J.F. Brown, Jr
IR Spectrum – J. Amer. Chem. Soc. 78, 4225
(1956) – R.D. Kross and V.A. Fassel
IR Spectrum – Aust. J. Chem. 11, 513
(1958) – L.K. Dyall and A.N. Hambly
Vibrational Analysis – Spectrochim. Acta
12, 305 (1958) – C. Garrigou-Lagrange,
J.M. Lebas and M.L. Josien
Vibrational Analysis – Spectrochim. Acta
15, 225 (1959) – J.M. Lebas, C. Garri-
gou-Lagrange and M.L. Josien
Vibrational Analysis – Z. Elektrochem. 64,
853 (1960) – H.W. Schrotter
Vibrational Analysis – Spectrochim. Acta
17, 523 (1961) – J. Brandmuller, E.W.
Schmid, H.W. Schrotter and G. Nonnen-
macher
IR Spectrum – Indian J. Phys. 37, 275
(1963) – K.C. Medhi and G.S. Kastha
IR Spectrum – Spectrochim. Acta 25A, 1423
(1969) – L.K. Dyall
IR Spectrum – Aust. J. Chem. 23, 947
(1970) – L.K. Dyall
Vibrational Analysis – Indian J. Pure Appl.
Phys. 8, 857 (1970) – S.C. Srivastava,
M.P. Srivastava and I.S. Singh
Fluorescence – J. Chem. Phys. 58, 1607
(1973) – O.S. Khalil, C.J. Seliskar
and S.P. McGlynn
UV Absorption in Vapour – Spectrochim. Acta
33A, 21 (1977) – S. Millefiori, G.
Favini, A. Millefiori and D. Grasso

o-Nitroanisole ($C_6H_4OCH_3NO_2$)

Raman – Z. Phys. Chem. B42, 315 (1943) –
H. Wittek
UV Absorption in Solution – Can. J. Chem.
38, 1837 (1960) – J.C. Dearden and
W.F. Forbes
Vibrational Analysis – Indian J. Pure Appl.
Phys. 8, 857 (1970) – S.C. Srivastava,
M.P. Srivastava and I.S. Singh
IR Spectrum – Indian J. Pure Appl. Phys.
9, 376 (1971) – A.N. Pandey and N.K.
Sanyal
UV Absorption in Vapour – Curr. Sci. 45,
650 (1976) – M.A. Shashidhar, D.K.
Despande and K.S. Rao

m-Nitroanisole ($C_6H_4OCH_3NO_2$)

Raman – Z. Phys. Chem. B42, 315 (1943) –
H. Wittek

UV Absorption in Solution – Can. J. Chem.
38, 1837 (1960) – J.C. Dearden and
W.F. Forbes
Vibrational Analysis – J. Chim. Phys.
63, 552 (1966) – C. Garrigou-Lagrange,
M. Chehata and J. Lascombe
Vibrational Analysis – Indian J. Pure
Appl. Phys. 8, 857 (1970) – S.C.
Srivastava, M.P. Srivastava and I.S.
Singh

p-Nitroanisole ($C_6H_4OCH_3NO_2$)

Raman – Z. Phys. Chem. B42, 315 (1943) –
H. Wittek
IR Spectrum – J. Amer. Chem. Soc. 77, 341
(1955) – J.F. Brown, Jr
IR Spectrum – J. Amer. Chem. Soc. 78,
4225 (1956) – R.D. Kross and V.A.
Fassel
Vibrational Analysis – Indian J. Pure
Appl. Phys. 8, 857 (1970) – S.C.
Srivastava, M.P. Srivastava and I.S.
Singh
IR Spectrum – Indian J. Pure Appl. Phys.
9, 376 (1971) – A.N. Pandey and N.K.
Sanyal
UV Absorption in Solution – Curr. Sci.
45, 650 (1976) – M.A. Shashidhar,
D.K. Despande and K.S. Rao

9-Nitroanthracene ($C_{14}H_9NO_2$)

UV Absorption in Solution – Indian J.
Pure Appl. Phys. 14, 53 (1976) –
K.M. Jain, B.N. Mali and T.N. Misra

o-Nitrobenzaldehyde ($C_6H_4CHONO_2$)

Raman – Z. Phys. Chem. B42, 315 (1943) –
H. Wittek
IR Spectrum – Ric. Sci. 24, 1687 (1954) –
M. Scrocco and A. Liberti
IR Spectrum – Anal. Chem. 27, 2 (1955) –
S. Pinchas
IR Spectrum – J. Chem. Soc. 333 (1959) –
A.R. Katritzky and P. Simons
Vibrational Analysis – Indian J. Pure
Appl. Phys. 7, 504 (1969) – M.P.
Srivastava, O.N. Singh and I.S. Singh
Calculation – J. Mol. Struct. 18, 457
(1973) – A.I. Kiss and J. Szoke

m-Nitrobenzaldehyde ($C_6H_4CHONO_2$)

Raman — Z. Phys. Chem. B42, 315 (1943) — H. Wittek

IR Spectrum — Ric. Sci. 24, 1687 (1954) — M. Scrocco and A. Liberti

IR Spectrum — J. Amer. Chem. Soc. 77, 6341 (1955) — J.F. Brown, Jr

IR Spectrum — Anal. Chem. 27, 2 (1955) — S. Pinchas

IR Spectrum — J. Chem. Soc. 333 (1959) — A.R. Katritzky and P. Simons

Vibrational Analysis — J. Chim. Phys. 63, 552 (1966) — C. Garrigou-Lagrange, M. Chehata and J. Lascombe

Vibrational Analysis — Indian J. Pure Appl. Phys. 7, 504 (1969) — M.P. Srivastava, O.N. Singh and I.S. Singh

Calculation — J. Mol. Struct. 18, 457 (1973) — A.I. Kiss and J. Szoke

p-Nitrobenzaldehyde ($C_6H_4CHONO_2$)

Raman — Z. Phys. Chem. B42, 315 (1943) — H. Wittek

IR Spectrum — Ric. Sci. 24, 1687 (1954) — M. Scrocco and A. Liberti

IR Spectrum — Anal. Chem. 27, 2 (1955) — S. Pinchas

IR Spectrum — J. Chem. Soc. 333 (1959) — A.R. Katritzky and P. Simons

Vibrational Analysis — Indian J. Pure Appl. Phys. 7, 504 (1969) — M.P. Srivastava, O.N. Singh and I.S. Singh

Calculation — J. Mol. Struct. 18, 457 (1973) — A.I. Kiss and J. Szoke

Nitrobenzene ($C_6H_5NO_2$)

Raman — Z. Phys. Chem. B42, 315 (1943) — H. Wittek

IR Spectrum — J. Amer. Chem. Soc. 77, 6341 (1955) — J.F. Brown, Jr

Vibrational Analysis — Z. Elektrochem. 62, 544 (1958) — J. Behringer

Vibrational Analysis — Z. Naturforsch. 159, 85 (1960) — K.E. Reinert

Vibrational Analysis — Spectrochim. Acta 17, 933 (1961) — C.V. Stephenson, W.C. Coburn, Jr and W.S. Wilcox

Vibrational Analysis — Spectrochim. Acta 17, 486 (1961) — J.H.S. Green, W. Kynaston and A.S. Lindsey

Vibrational Analysis — J. Chim. Phys. 59, 1072 (1962) — J.M. Lebas

Vibrational Analysis — Spectrochim. Acta 18, 561 (1962) — S. Dachne and H. Stanke

Vibrational Analysis — J. Chim. Phys. 61, 1439 (1964) — P. Delorme

UV Absorption in Solution — J. Mol. Spectrosc. 13, 174 (1964) — S. Nagakura, M. Kojima and Y. Marurama

IR Spectrum — Spectrochim. Acta 22, 501 (1966) — W.R. McWhinnie and R.C. Poller

Vibrational Analysis — Spectrochim. Acta 23A, 728 (1967) — V.C. Former

Vibrational Analysis — Proc. Roy. Soc. A298, 51 (1967) — P.R. Griffiths and H.W. Thompson

Vibrational Analysis — Spectrochim. Acta 26A, 1925 (1970) — J.H.S. Green and D.J. Harrison

Vibrational Analysis — Spectrochim. Acta 26A, 1515 (1970) — J.H.S. Green, D.J. Harrison and D.W. Scott

Phosphorescence — J. Mol. Spectrosc. 35, 455 (1970) — O.S. Khalil, H.G. Bach and S.P. McGlynn

Calculation — J. Mol. Struct. 16, 365 (1973) — R.T.C. Brownlee, D.G. Cameron, R.D. Topsom, A.R. Katritzky and A.J. Sparrow

Calculation — J. Mol. Struct. 18, 457 (1973) — A.I. Kiss and J. Szoke

UV Absorption in Vapour — J. Chem. Soc. Faraday II 70, 1761 (1974) — V.V. Bhujle and C.N.R. Rao

Raman (Inverse) — Chem. Phys. Lett. 29, 389 (1974) — S.H. Lin, E.S. Reid and C.J. Tredwell

Nitrobenzene-p-D_1 ($C_6H_4DNO_2$)

Vibrational Analysis — Spectrochim. Acta 35A, 27 (1979) — A. Kuwae and K. Machida

Nitrobenzene-D_5 ($C_6D_5NO_2$)

Vibrational Analysis — Spectrochim. Acta 35A, 27 (1979) — A. Kuwae and K. Machida

Vibrational Analysis — Spectrochim. Acta 35A, 165 (1979) — J.D. Laposa

m-Nitrobenzonitrile ($C_6H_4CNNO_2$)

Vibrational Analysis — J. Chim. Phys. 63, 552 (1966) — C. Garrigou-Lagrange, M. Chehata and J. Lascombe

p-Nitrobenzoylchloride $(C_6H_4NO_2COCl)$

Vibrational Analysis – Spectrochim. Acta
  $\underline{12}$, 305 (1958) – C. Garrigou-Lagrange,
  J.M. Lebas and M.L. Josien
Vibrational Analysis – Spectrochim. Acta
  $\underline{15}$, 225 (1959) – J.M. Lebas, C.
  Garrigou-Lagrange and M.L. Josien

p-Nitrobenzylbromide $(C_6H_4NO_2CH_2Br)$

Absorption Spectrum – Spectrochim. Acta
  $\underline{31A}$, 199 (1975) – R.M. Issa, A.M.G.
  Nassar, A.M. Hindawey, F.M. Issa and
  G.B. El-Hefnawey

p-Nitrobenzylchloride $(C_6H_4NO_2CH_2Cl)$

Absorption Spectrum – Spectrochim. Acta
  $\underline{31A}$, 199 (1975) – R.M. Issa, A.M.G.
  Nassar, A.M. Hindawey, F.M. Issa and
  G.B. El-Hefnawey

p-Nitrobenzylcyanide $(C_6H_4NO_2CH_2CN)$

Absorption Spectrum – Spectrochim. Acta
  $\underline{31A}$, 199 (1975) – R.M. Issa, A.M.G.
  Nassar, A.M. Hindawey, F.M. Issa and
  G.B. El-Hefnawey

p-Nitrobenzyliodide $(C_6H_4NO_2CH_2I)$

Absorption Spectrum – Spectrochim. Acta
  $\underline{31A}$, 199 (1975) – R.M. Issa, A.M.G.
  Nassar, A.M. Hindawey, F.M. Issa and
  G.B. El-Hefnawey

α-Nitronaphthalene $(C_{10}H_7NO_2)$

Fluorescence – J. Amer. Chem. Soc. $\underline{66}$, 2100
  (1944) – G.N. Lewis and M. Kasha
Fluorescence – J. Chem. Phys. $\underline{17}$, 905
  (1949) – D.S. McClure
Fluorescence – J. Mol. Spectrosc. $\underline{35}$, 455
  (1970) – O.S. Khalil, H.G. Bach and
  S.P. McGlynn
Phosphorescence – J. Mol. Spectrosc. $\underline{35}$,
  455 (1970) – O.S. Khalil, H.G. Bach
  and S.P. McGlynn
Phosphorescence – Spectrochim. Acta $\underline{27A}$,
  787 (1971) – R. Rusakowicz and A.C.
  Testa

Fluorescence – J. Chem. Soc. Faraday II
  $\underline{70}$, 1159 (1974) – C. Capellos and G.
  Porter

β-Nitronaphthalene $(C_{10}H_7NO_2)$

Phosphorescence – Spectrochim. Acta $\underline{27A}$,
  787 (1971) – R. Rusakowicz and A.C.
  Testa

o-Nitrophenetole $(C_6H_4OC_2H_5NO_2)$

UV Absorption in Different States – Indian
  J. Phys. $\underline{34}$, 289 (1960) – K.V.K. Rao
  and V.R. Rao
IR Spectrum – Indian J. Pure Appl. Phys.
  $\underline{9}$, 756 (1971) – K.M. Mathur, D.P.
  Duyal and R.N. Singh

p-Nitrophenetole $(C_6H_4OC_2H_5NO_2)$

UV Absorption in Different States – Indian
  J. Phys. $\underline{34}$, 289 (1960) – K.V.K. Rao
  and V.R. Rao
IR Spectrum – Indian J. Pure Appl. Phys.
  $\underline{9}$, 756 (1971) – K.M. Mathur, D.P.
  Juyal and R.N. Singh

o-Nitrophenol $(C_6H_4OHNO_2)$

Raman – Monatsh. Chem. $\underline{67}$, 92 (1936) –
  A.W. Reitz and W. Stockmair
IR Spectrum – Z. Elektrochem. $\underline{60}$, 136
  (1956) – V. Keussler and G. Rossmi
Raman – Z. Elektrochem. $\underline{62}$, 544 (1958) –
  J. Behringer
UV Absorption in Solution – Can. J. Chem.
  $\underline{38}$, 1837 (1960) – J.C. Dearden and
  W.F. Forbes
Vibrational Analysis – Spectrochim. Acta
  $\underline{17}$, 486 (1961) – J.H.S. Green, W.
  Kynaston and A.S. Lindsey
IR Spectrum – Indian J. Phys. $\underline{36}$, 163
  (1962) – S.B. Banerjee and G.S. Kastha
IR Spectrum – Opt. Spektrosk. $\underline{16}$, 78
  (1964) – A.E. Stanevich
IR Spectrum – Indian J. Phys. $\underline{48}$, 412
  (1974) – Y. Kishore, S.N. Sharma and
  C.P.D. Dwivedi

m-Nitrophenol ($C_6H_4OHNO_2$)

UV Absorption in Solution — Can. J. Chem.
  <u>38</u>, 1837 (1960) — J.C. Dearden and W.F.
  Forbes
IR Spectrum — Opt. Spektrosk. <u>16</u>, 781
  (1964) — A.E. Stanevich
Vibrational Analysis — J. Chim. Phys. <u>63</u>,
  552 (1966) — C. Garrigou-Lagrange, M.
  Chehata and J. Lascombe
IR Spectrum — Indian J. Phys. <u>48</u>, 412
  (1974) — Y. Kishore, S.N. Sharma and
  C.P.D. Dwivedi

p-Nitrophenol ($C_6H_4OHNO_2$)

IR Spectrum — Naturwissenschaften <u>41</u>, 333
  (1954) — Y. Ginnetti
IR Spectrum — J. Amer. Chem. Soc. <u>77</u>, 6341
  (1955) — J.F. Brown, Jr
IR Spectrum — J. Amer. Chem. Soc. <u>78</u>, 4225
  (1956) — R.D. Kross and V.A. Fassel
Vibrational Analysis — Spectrochim. Acta
  <u>12</u>, 305 (1958) — C. Garrigou-Lagrange,
  J.M. Lebas and M.L. Josien
Vibrational Analysis — Spectrochim. Acta
  <u>15</u>, 225 (1959) — J.M. Lbas, C.
  Garrigou-Lagrange and M.L. Josien
IR Spectrum — Spectrochim. Acta <u>16</u>, 58
  (1960) — E.R. Lippincott, C.E. Weir,
  A. van Valkenberg and E.N. Bunting
IR Spectrum — Spectrochim. Acta <u>17</u>, 1275
  (1961) — J. Rohlender
IR Spectrum — Indian J. Phys. <u>36</u>, 163
  (1962) — S.B. Banerjee and G.S. Kastha
IR Spectrum — Appl. Spectrosc. <u>16</u>, 32
  (1962) — R.J. Jackobsen and E.J. Brewer
IR Spectrum — Opt. Spektrosk. <u>16</u>, 781
  (1964) — A.E. Stanevich
IR Spectrum — Indian J. Phys. <u>48</u>, 412
  (1974) — Y. Kishore, S.N. Sharma and
  C.P.D. Dwivedi

o-Nitropyridine ($C_5H_4NNO_2$)

UV Absorption in Solution — Spectrochim.
  Acta <u>23A</u>, 89 (1967) — G. Favini, A.
  Gambo and I.R. Bellobono

m-Nitropyridine ($C_5H_4NNO_2$)

UV Absorption in Solution — Spectrochim.
  Acta <u>23A</u>, 89 (1967) — G. Favini, A.
  Gambo and I.R. Bellobono

p-Nitropyridine ($C_5H_4NNO_2$)

UV Absorption in Solution — Spectrochim.
  Acta <u>23A</u>, 89 (1967) — G. Favini, A.
  Gambo and I.R. Bellobono

m-Nitrostyrene ($C_6H_4CH:CH_2NO_2$)

Vibrational Analysis — Appl. Spectrosc.
  <u>22</u>, 650 (1968) — W.G. Fateley, G.L.
  Carlson and F.E. Dickson

o-Nitrotoluene ($C_6H_4CH_3NO_2$)

Vibrational Analysis — Spectrochim. Acta
  <u>12</u>, 305 (1958) — C. Garrigou-Lagrange,
  J.M. Lebas and M.L. Josien
Vibrational Analysis — Spectrochim. Acta
  <u>26A</u>, 1925 (1970) — J.H.S. Green and
  D.J. Harrison
Raman — J. Mol. Struct. <u>5</u>, 477 (1970) —
  N.T. McDevitt and W.G. Fateley

m-Nitrotoluene ($C_6H_4CH_3NO_2$)

IR Spectrum — J. Amer. Chem. Soc. <u>77</u>, 6341
  (1955) — J.F. Brown, Jr
Raman — Z. Elektrochem. <u>62</u>, 544 (1958) —
  J. Behringer
Vibrational Analysis — J. Chim. Phys. <u>69</u>,
  552 (1966) — C. Garrigou-Lagrange, M.
  Chehata and J. Lascombe
Vibrational Analysis — Spectrochim. Acta
  <u>26A</u>, 1925 (1970) — J.H.S. Green and
  D.J. Harrison

p-Nitrotoluene ($C_6H_4CH_3NO_2$)

IR Spectrum — J. Amer. Chem. Soc. <u>77</u>, 6341
  (1955) — J.F. Brown, Jr
IR Spectrum — J. Amer. Chem. Soc. <u>78</u>, 4225
  (1956) — R.D. Kross and V.A. Fassel
Vibrational Analysis — Spectrochim. Acta
  <u>12</u>, 305 (1958) — C. Garrigou-Lagrange,
  J.M. Lebas and M.L. Josien
Vibrational Analysis — Bull. Soc. Chim.
  Fr. 93 (1959) — N. Fuson, M.L. Josien,
  J. Deschamps, C. Garrigou-Lagrange
  and M.J. Forel
Vibrational Analysis — Spectrochim. Acta
  <u>15</u>, 225 (1959) — J.M. Lebas, C.
  Garrigou-Lagrange and M.L. Josien
Vibrational Analysis — Spectrochim. Acta
  <u>26A</u>, 1925 (1970) — J.H.S. Green and
  D.J. Harrison

Octofluorotoluene ($C_6F_5CF_3$)

Vibrational Analysis — Spectrochim. Acta
24A, 1891 (1968) — R.T. Bailey and
S.G. Hasson

Vibrational Analysis — Spectrochim. Acta
30A, 1225 (1974) — S.G. Frankiss, D.J.
Harrison and W. Kynaston

Pentabromotoluene ($C_6Br_5CH_3$)

IR Spectrum — Spectrochim. Acta 28A, 103
(1972) — A.B. Dempster, D.B. Powell and
N. Sheppard

Pentachlorobenzene (C_6Cl_5H)

Vibrational Analysis — Spectrochim. Acta
19, 1739 (1963) — J.R. Scherer and
J.C. Evans

Pentachlorobenzene-D_1 (C_6Cl_5D)

Vibrational Analysis — Spectrochim. Acta
19, 1739 (1963) — J.R. Scherer and J.C.
Evans

Pentachlorophenol (C_6Cl_5OH)

Vibrational Analysis — J. Mol. Struct. 33,
307 (1976) — J.H.S. Green, D.J. Harrison
and C.P. Stockley

Pentachloropyridine (C_5Cl_5N)

Vibrational Analysis — Spectrochim. Acta
26A, 1129 (1970) — R.T. Bailey and
G.P. Strachan

Pentachlorothiophenol (C_6Cl_5SH)

Vibrational Analysis — J. Mol. Struct. 33,
307 (1976) — J.H.S. Green, D.J.
Harrison and C.P. Stockley

Pentachlorotoluene ($C_6Cl_5CH_3$)

IR Spectrum — Spectrochim. Acta 28A, 103
(1972) — A.B. Dempster, D.B. Powell
and N. Sheppard

Pentafluoroanisole ($C_6F_5OCH_3$)

IR Spectrum — Indian J. Pure Appl. Phys.
7, 519 (1969) — R. Amni Amma, K.P.R.
Nair and D.K. Rai

Pentafluorobenzene (C_6F_5H)

Calculation — Spectrochim. Acta. 19, 1947
(1947) — D.A. Long and D. Steele

IR Spectrum — Chem. Ind.(London) 821
(1957) — R. Stephens and J.C. Tatlow

Vibrational Analysis — Spectrochim. Acta
16, 368 (1960) — D. Steele and D.H.
Whiffen

Vibrational Analysis — Proc. Indian Acad.
Sci. A68, 53 (1968) — S. Doraiswamy
and S.D. Sharma

Fluorescence — Chem. Phys. Lett. 13, 140
(1972) — G.L. Loper and E.K.C. Lee

Calculation — J. Mol. Spectrosc. 48, 446
(1973) — V.J. Eaton and D. Steele

Vibrational Analysis — Spectrochim. Acta
31A, 1839 (1975) — S.G. Frankiss and
D.J. Harrison

Calculation — J. Mol. Struct. 32, 93
(1976) — F. Torok, A. Hegedus, K. Kosa
and P. Puley

Pentafluorobenzene-D_1 (C_6F_5D)

Vibrational Analysis — Spectrochim. Acta
16, 368 (1960) — D. Steele and D.H.
Whiffen

Calculation — Spectrochim. Acta 19, 1947
(1963) — D.A. Long and D. Steele

Pentafluorobenzenethiol (C_6F_5SH)

Vibrational Analysis — Spectrochim. Acta
33A, 423 (1977) — J.H.S. Green, D.J.
Harrison and C.P. Stockley

Pentafluorobenzonitrile (C_6F_5CN)

Vibrational Analysis — Spectrochim. Acta
24A, 1257 (1968) — H.F. Shurvell, A.S.
Blair and R.J. Jackobsen

Vibrational Analysis — Spectrochim. Acta
32A, 1195 (1976) — J.H.S. Green and
D.J. Harrison

Pentafluorobenzoylchloride (C_6F_5COCl)

Vibrational Analysis — Curr. Sci. <u>41</u>, 471
(1972) — K. Singh, S.R. Singh and I.S.
Singh
UV Absorption in Solution — Indian J. Pure
Appl. Phys. <u>11</u>, 512 (1973) — K. Singh,
B.B. Lal and I.S. Singh

Pentafluorobenzodifluoride ($C_6F_5CF_2$)

Vibrational Analysis — Spectrochim. Acta
<u>24A</u>, 1999 (1968) — E.F. Mooney

Pentafluorobenzotrifluoride ($C_6F_5CF_3$)

Vibrational Analysis — Spectrochim. Acta
<u>25A</u>, 467 (1969) — R.T. Bailey and
S.S. Hasson

Pentafluorobenzylchloride ($C_6F_5CH_2Cl$)

Vibrational Analysis — Spectrochim. Acta
<u>24A</u>, 1999 (1968) — E.F. Mooney

Pentafluorobromobenzene (C_6F_5Br)

Vibrational Analysis — Spectrochim. Acta
<u>19</u>, 1955 (1963) — D.A. Long and D.
Steele
Vibrational Analysis — Spectrochim. Acta
<u>22</u>, 695 (1966) — I.J. Hyams, E.R.
Lippincott and R.T. Bailey

Pentafluorochlorobenzene (C_6F_5Cl)

Vibrational Analysis — Spectrochim. Acta
<u>19</u>, 1955 (1963) — D.A. Long and D.
Steele
Vibrational Analysis — Spectrochim. Acta
<u>22</u>, 695 (1966) — I.J. Hyams, E.R.
Lippincott and R.T. Bailey
Vibrational Analysis — Spectrochim. Acta
<u>31A</u>, 1839 (1975) — S.G. Frankiss and
D.J. Harrison

Pentafluoroiodobenzene (C_6F_5I)

Vibrational Analysis — Spectrochim. Acta
<u>19</u>, 1955 (1963) — D.A. Long and D.
Steele

Vibrational Analysis — Spectrochim. Acta
<u>22</u>, 695 (1966) — I.J. Hyams, E.R.
Lippincott and R.T. Bailey
Vibrational Analysis — Spectrochim. Acta
<u>31A</u>, 1839 (1975) — S.G. Frankiss and
D.J. Harrison
Vibrational Analysis — Spectrochim. Acta
<u>33A</u>, 423 (1977) — J.H.S. Green, D.J.
Harrison and C.P. Stockley

Pentafluoronitrobenzene ($C_6F_5NO_2$)

Vibrational Analysis — Spectrochim. Acta
<u>33A</u>, 423 (1977) — J.H.S. Green, D.J.
Harrison and C.P. Stockley

Pentafluorophenol (C_6F_5OH)

UV Absorption in Solution — Indian J. Pure
Appl. Phys. <u>11</u>, 512 (1973) — K. Singh,
B.B. Lal and I.S. Singh

Pentafluoropyridine (C_5F_5N)

Vibrational Analysis — Trans. Faraday Soc.
<u>59</u>, 599 (1963) — D.A. Long and R.T.
Bailey
Vibrational Analysis — Spectrochim. Acta
<u>33A</u>, 81 (1977) — J.H.S. Green and
D.J. Harrison

Pentafluorostyrene ($C_6F_5CH:CH_2$)

Vibrational Analysis — Curr. Sci. <u>41</u>, 133
(1972) — K. Singh and I.S. Singh
Vibrational Analysis — Spectrochim. Acta
<u>33A</u>, 423 (1977) — J.H.S. Green, D.J.
Harrison and C.P. Stockley
Vibrational Analysis — Spectrochim. Acta
<u>33A</u>, 723 (1977) — M. Towland, A.M.
North and R.A. Pethrick

Pentafluorotoluene ($C_6F_5CH_3$)

Calculation — Spectrochim. Acta <u>19</u>, 1947
(1963) — D.A. Long and D. Steele
Vibrational Analysis — Spectrochim. Acta
<u>24A</u>, 1891 (1968) — R.T. Bailey and
S.G. Hasson
IR Spectrum — Indian J. Pure Appl. Phys.
<u>7</u>, 519 (1969) — R. Amni Amma, K.P.R.
Nair and D.K. Rai

Vibrational Analysis — Spectrochim. Acta
25A, 467 (1969) — R.T. Bailey and
S.G. Hasson
IR Spectrum — Spectrochim. Acta 28A, 103
(1972) — A.B. Dempster, D.B. Powell
and N. Sheppard
Vibrational Analysis — Spectrochim. Acta
31A, 1839 (1975) — S.G. Frankiss and
D.J. Harrison

Pentafluorothiophenol (C_6F_5SH)

Vibrational Analysis — Curr. Sci. 41, 199
(1972) — K. Singh, S.C. Srivastava
and I.S. Singh

Pentamethylbenzene (C_6H(CH_3)_5)

IR Spectrum — Spectrochim. Acta 4, 373
(1951) — C.C. Cannon and G.B.B.M.
Sutherland
IR Spectrum — J. Phys. Chem. 61, 730
(1957) — S.H. Hestings and D.E.
Nicholson

Phenanthrene (C_14H_10)

Phosphorescence — J. Chem. Phys. 17, 905
(1949) — D.S. McClure
Phosphorescence — Bull. Chem. Soc. Jap.
27, 189 (1954) — S. Kato and M. Koizumi
Absorption Spectrum — J. Chem. Phys. 25,
481 (1956) — D.S. McClure
Absorption Spectrum — Opt. Spektrosk. 5,
582 (1958) — A.F. Prikhot'ko and A.
Fugol
UV Absorption in Solution — Spectrochim.
Acta 16, 1060 (1960) — R.N. Jones and
E. Spinner
Absorption Spectrum — J. Chim. Phys. 59,
167 (1962) — P. Pesteil and M. Rabauld
Phosphorescence — J. Chem. Phys. 39, 1186
(1963) — T. Azumi and S.P. McGlynn
Absorption Spectrum — Proc. Roy. Soc. A288,
69 (1965) — D.P. Craig and R.D. Gordon
Fluorescence — Proc. Roy. Soc. A288, 69
(1965) — D.P. Craig and R.D. Gordon
Fluorescence — J. Chem. Phys. 45, 2270
(1966) — R.M. Hochstrasser and G.L.
Small
Vibrational Analysis — J. Chem. Phys. 44,
2724 (1966) — V. Schettno, N. Netto and
S. Califano
Phosphorescence — Chem. Rev. 66, 199 (1966)—
S.K. Lower and M.A. El-Sayed

Absorption Spectrum (High Resolution) —
J. Mol. Spectrosc. 23, 219 (1967) —
G.A. Gerhold
Phosphorescence — J. Mol. Spectrosc. 27,
450 (1968) — M. Nakamize and T.
Matsueda

1,2-Phenanthrenequinone (C_14H_8O_2)

Vibrational Analysis — Curr. Sci. 39, 405
(1970) — G.D. Baruah, D.N. Tripathi
and S. Nath Singh

Phenazine (C_12H_8N_2)

IR Spectrum — Spectrochim. Acta 19, 1625
(1963) — C. Stammer and A. Taurins
IR Spectrum — Spectrochim. Acta 20, 1503
(1964) — N. Netto, F. Ambrosine and
S. Califano
Raman — J. Chem. Phys. 51, 2936 (1969) —
T.G. Pavlopoulos
Raman — J. Mol. Spectrosc. 39, 536
(1971) — T.J. Durnick and S.C. Wait,Jr
Vibrational Analysis — J. Mol. Spectrosc.
42, 211 (1972) — T.J. Durnick and S.C.
Wait, Jr

p-Phenetidine (C_6H_4NH_2OC_2H_5)

UV Absorption in Vapour — Indian J. Pure
Appl. Phys. 5, 307 (1967) — P.D. Singh
UV Absorption in Vapour — Indian J. Pure
Appl. Phys. 7, 518 (1969) — R. Amni
Amma, G.D. Baruah and K.P.R. Nair
IR Spectrum — Indian J. Pure Appl. Phys.
10, 406 (1972) — B.B. Lal, M.P. Sriva-
stava and I.S. Singh

Phenetole (C_6H_5OC_2H_5)

UV Absorption in Vapour — J. Amer. Chem.
Soc. 72, 1539 (1950) — W.W. Robertson,
A.J. Sariff and F.A. Matsen
UV Absorption in Vapour — Indian J. Phys.
25, 123 (1951) — K. Sreeramamurty
Raman — Indian J. Phys. 30, 530 (1956) —
D.C. Biswas
Vibrational Analysis — Spectrochim. Acta
18, 39 (1962) — J.H.S. Green

104

Phenol (C_6H_5OH)

Vibrational Analysis – Proc. Indian Acad. Sci. 3A, 52 (1936) – R. Ananthkrishnan
UV Absorption in Vapour – J. Chem. Phys. 13, 167 (1945) – N. Ginsburg and F.A. Matsen
UV Absorption in Vapour – J. Chem. Phys. 13, 309 (1945) – F.A. Matsen, N. Ginsburg and W.W. Robertson
UV Absorption in Solution – J. Chem. Phys. 16, 274 (1948) – M. Davies
Phosphorescence – J. Chem. Phys. 17, 905 (1949) – D.S. McClure
UV Absorption in Vapour – J. Amer. Chem. Soc. 72, 1539 (1950) – W.W. Robertson, A.J. Sariff and F.A. Matsen
UV Absorption in Vapour – J. Chem. Phys. 18, 1135 (1950) – C.A. Beck
UV Absorption in Solution – Indian J. Phys. 30, 353 (1956) – S.B. Banerjee
IR Spectrum – Indian J. Phys. 32, 345 (1958) – S.C. Sirkar, A.R. Deb and S.B. Banerjee
Vibrational Analysis – Spectrochim. Acta 16, 1382 (1960) – J.C. Evans
IR Spectrum – Spectrochim. Acta 16, 528 (1960) – A. Hidalgo and C. Otero
Vibrational Analysis – J. Chem. Soc. 2236 (1961) – J.H.S. Green
UV Emission in Vapour – Proc. Natl. Acad. Sci. (India) 32A, 302 (1962) – N.L. Singh
Vibrational Analysis – Spectrochim. Acta 18, 39 (1962) – J.H.S. Green
Fluorescence – Indian J. Pure Appl. Phys. 3, 374 (1965) – S. Prakash and N.L. Singh
Vibrational Analysis – Opt. Spektrosk. 18, 152 (1965) – M.A. Kovner, N.I. Davydova and I.A. Zhigunova
IR Spectrum – Spectrochim. Acta 22, 501 (1966) – W.R. McWhinnie and R.C. Poller
Vibrational Analysis – J. Chem. Phys. 45, 1736 (1966) – H. Forest and D.P. Dailey
UV Absorption in Vapour (High Resolution) – J. Mol. Spectrosc. 21, 76 (1966) – H.D. Bist, J.C.D. Brand and D.R. Williams
UV Absorption in Vapour (High Resolution) – J. Mol. Spectrosc. 24, 413 (1967) – H.D. Bist, J.C.D. Brand and D.R. Williams
UV Absorption in Vapour (High Resolution) – Proc. Roy. Soc. A307, 97 (1968) – J. Christoffersen, J.M. Hollas and G.H. Kirby
Calculation – J. Mol. Struct. 16, 365 (1973) – R.T.C. Brownlee, D.G. Cameron, R.D. Topsom, A.R. Katritzky and A.J. Sparrow

IR Spectrum – J. Mol. Struct. 22, 29 (1974) – N.W. Larsen and F.M. Nicolaisen

Phenol-OD (C_6H_5OD)

Vibrational Analysis – Z. Elektrochem. 59, 866 (1955) – R. Mecke and G. Rossmy
Vibrational Analysis – Spectrochim. Acta 16, 1382 (1960) – J.C. Evans
Vibrational Analysis – J. Mol. Spectrosc. 24, 402 (1967) – H.D. Bist, J.C.D. Brand and D.R. Williams

Phenol-D_4 (C_6HD_4OH)

IR Spectrum – J. Mol. Struct. 22, 29 (1974) – N.W. Larsen and F.M. Nicolaisen

Phenol-D_5 (C_6D_5OH)

Vibrational Analysis – Spectrochim. Acta 16, 1382 (1960) – J.C. Evans
IR Spectrum – J. Mol. Spectrosc. 24, 402 (1967) – H.D. Bist, J.C.D. Brand and D.R. Williams

Phenol-D_6 (C_6D_5OD)

IR Spectrum – J. Mol. Struct. 22, 29 (1974) – N.W. Larsen and F.M. Nicolaisen

o-Phenylenediamine ($C_6H_4(NH_2)_2$)

IR Spectrum – J. Phys. Radium 8, 489 (1937) – J. Lecomte
IR Spectrum – J. Chem. Phys. 6, 202 (1938) – O.R. Wolf and L.S. Daming
Raman – Monatsh. Chem. 72, 244 (1939) – O. Paulsen
Raman – Monatsh. Chem. 72, 378 (1939) – E. Pandel and G. Ralinger
Raman – Monatsh. Chem. 74, 175 (1943) – E. Herz and K.W.F. Kohlrausch
UV Absorption in Vapour – J. Chem. Phys. 23, 1740 (1955) – A. Sado and T. Anno
Calculation – J. Mol. Struct. 6, 246 (1970) – P.C. Mishra and D.K. Rai
Calculation – J. Mol. Struct. 13, 253 (1972) – J.S. Yadava, P.C. Mishra and D.K. Rai

m-Phenylenediamine ($C_6H_4(NH_2)_2$)

IR Spectrum – J. Phys. Radium 8, 489
    (1937) – J. Lecomte
IR Spectrum – J. Chem. Phys. 6, 202
    (1938) – O.R. Wolf and L.S. Daming
Raman – Monatsh. Chem. 72, 244 (1939) –
    O. Paulsen
Raman – Monatsh. Chem. 72, 378 (1939) – E.
    Pandel and G. Radinger
Raman – Monatsh. Chem. 74, 175 (1943) –
    E. Herz and K.W.F. Kohlrausch
UV Absorption in Vapour – J. Chem. Phys.
    23, 1740 (1955) – A. Sado and T. Anno
IR Spectrum – Proc. Roy. Soc. A243, 143
    (1957) – P.J. Krueger and H.W. Thompson
Calculation – J. Mol. Struct. 6, 246
    (1970) – P.C. Mishra and D.K. Rai

p-Phenylenediamine ($C_6H_4(NH_2)_2$)

IR Spectrum – J. Phys. Radium 8, 489
    (1937) – J. Lecomte
IR Spectrum – J. Chem. Phys. 6, 202
    (1938) – O.R. Wolf and L.S. Daming
Raman – Monatsh. Chem. 72, 244 (1939) –
    O. Paulsen
Raman – Monatsh. Chem. 72, 378 (1939) –
    E. Pandel and G. Radinger
Raman – Monatsh. Chem. 74, 175 (1943) –
    E. Herz and K.W.F. Kohlrausch
UV Absorption in Vapour – J. Chem. Phys.
    23, 1740 (1955) – A. Sado and T. Anno
IR Spectrum – Proc. Roy. Soc. A243, 143
    (1957) – P.J. Krueger and H.W. Thompson
Calculation – J. Mol. Struct. 6, 246
    (1970) – P.C. Mishra and D.K. Rai
Calculation – J. Mol. Struct. 13, 253
    (1972) – J.S. Yadava, P.C. Mishra and
    D.K. Rai
Calculation – J. Mol. Struct. 14, 61
    (1972) – L. Smetankine and J. Etchepare

4-Phenylbenzophenone ($C_{19}H_{14}O$)

Phosphorescence – Opt. Spektrosk. 24, 607
    (1970) – A.A. Kotov

Phthalaldehyde ($C_6H_4(CHO)_2$)

IR Spectrum – Indian J. Pure Appl. Phys.
    9, 857 (1971) – B.B. Lal, M.P.
    Srivastava and I.S. Singh

Phthalazine ($C_8H_6N_2$)

Vibrational Analysis – J. Mol. Spectrosc.
    36, 310 (1970) – R.W. Mitchell, R.W.
    Glass and J.A. Merritt
Calculation – J. Mol. Spectrosc. 44, 118
    (1972) – H. Baba and I. Yamazaki
Calculation – J. Mol. Spectrosc. 50, 457
    (1974) – J.E. Ridley

Picoline ($C_5H_4NCH_3$)

See   Methylpyridine

Piperazine ($C_4H_4(NH)_2$)

Vibrational Analysis – Spectrochim. Acta
    18, 299 (1962) – P.J. Hendra and D.B.
    Powell

Piperidine ($C_5H_{11}N$)

UV Absorption in Vapour – Indian J. Pure
    Appl. Phys. 10, 490 (1972) – S. Nath
    Singh, G.D. Baruah and R.S. Singh

Pseudocumidine ($C_6H_2NH_2(CH_3)_3$)

UV Absorption in Solution – Bull. Soc.
    Chim. Fr. 45, 134 (1949) – P.
    Gramanaticakis
IR Spectrum – Proc. Natl. Dakata. Akad.
    Sci. 12, 33 (1958) – E.J. Kaatz and
    E.J. O'Reilly
Vibrational Analysis – Indian J. Pure
    Appl. Phys. 8, 678 (1970) – G.D.
    Baruah, K. Singh and D.K. Rai

Pyrazine ($C_4H_4N_2$)

IR Spectrum – J. Chem. Phys. 25, 597
    (1956) – M. Ito, R. Shimada, T. Kura-
    ishi and W. Mizushima
UV Absorption in Vapour – J. Chem. Phys.
    26, 1508 (1957) – M. Ito, R. Shimada,
    T. Kuraishi and W. Mizushima
Vibrational Analysis – Spectrochim. Acta
    9, 113 (1957) – R.C. Lord, A.L.
    Marston and F.A. Miller
UV Absorption in Solution – Spectrochim.
    Acta 12, 114 (1958) – R.C. Hirt
Phosphorescence – J. Mol. Spectrosc. 2,
    58 (1958) – L. Goodman and M. Kasha

UV Absorption in Vapour - Spectrochim. Acta 16, 945 (1960) - J. Merritt and K.K. Innes

Calculation - Spectrochim. Acta 17, 233 (1961) - D.B. Scully

Phosphorescence - Spectrochim. Acta 17, 14 (1961) - R. Shimada

UV Absorption in Vapour (High Resolution) - J. Mol. Spectrosc. 11, 257 (1963) - K.K. Innes, J.D. Simmons and S.G. Tilford

UV Absorption in Vapour (High Resolution) - Disc. Faraday Soc. 35, 237 (1963) - K.K. Innes and L. Giddings

Vibrational Analysis - Spectrochim. Acta 20, 385 (1964) - S. Califano, G. Adembi and G. Sbrana

Vibrational Analysis - J. Mol. Spectrosc. 14, 190 (1964) - J.D. Simmons, K.K. Innes and G.M. Begun

Vibrational Analysis - Spectrochim. Acta 21, 571 (1965) - M. Scrocco, C. Di Lavre and S. Califano

UV Absorption Spectrum - J. Mol. Spectrosc. 18, 372 (1965) - B.S. Snowden, Jr and W.H. Eberhardt

UV Absorption in Vapour - J. Mol. Spectrosc. 15, 407 (1965) - J.E. Parkin and K.K. Innes

Raman (Crystal) - J. Chem. Phys. 44, 1001 (1966) - M. Ito and T. Shigeoka

UV Absorption in Vapour (High Resolution) - J. Mol. Spectrosc. 21, 66 (1966) - K.K. Innes and J.E. Parkin

IR Spectrum - J. Mol. Spectrosc. 21, 217 (1966) - H. Takahashi, K. Mamala and E.K. Plyer

UV Absorption Spectrum - J. Mol. Spectrosc. 22, 125 (1967) - K.K. Innes, J.P. Bryne and I.G. Ross

UV Absorption in Vapour (^{15}N) - J. Mol. Spectrosc. 23, 280 (1967) - K.K. Innes and J.A. Merritt

Fluorescence - J. Chem. Phys. 46, 713 (1967) - B.J. Cohen and L. Goodman

Fluorescence - Acta Phys. Pol. 34, 721 (1968) - L.M. Logan and I.G. Ross

IR Spectrum - J. Chem. Phys. 49, 955 (1968) - X. Gerbaux and A. Hadni

UV Absorption in Vapour - J. Mol. Spectrosc. 43, 477 (1972) - K.K. Innes, A.H. Kalantar, A.Y. Khan and T.J. Durnick

Phosphorescence - J. Mol. Spectrosc. 43, 239 (1972) - C.J. Marzzacco and E.F. Zalewski

Fluorescence - J. Mol. Spectrosc. 43, 239 (1972) - C.J. Marzzacco and E.F. Zalewski

Absorption Spectrum in Vaccum - J. Mol. Spectrosc. 44, 1 (1972) - R. Scheps, D. Florida and S.A. Rice

Absorption Spectrum - J. Mol. Spectrosc. 52, 21 (1974) - I. Suzuka, N. Mikami and M. Ito

Fluorescence - J. Mol. Spectrosc. 52, 21 (1974) - I. Suzuki, N. Mikami and M. Ito

Fluorescence - J. Mol. Spectrosc. 52, 1 (1974) - H.K. Hong and G.W. Robinson

UV Absorption in Vapour (High Resolution)- J. Mol. Spectrosc. 52, 130 (1974) - S.N. Thakur and K.K. Innes

Raman - J. Mol. Spectrosc. 60, 277 (1976)- K. Kamagawa and M. Ito

Fluorescence - J. Mol. Spectrosc. 60, 277 (1976) - K. Kamagawa and M. Ito

UV Absorption Spectrum - J. Mol. Spectrosc. 60, 277 (1976) - K. Kamagawa and M. Ito

Pyrazine-D$_4$ (C$_4$D$_4$N$_2$)

UV Absorption in Vapour - J. Mol. Spectrosc. 11, 257 (1963) - K.K. Innes, J.D. Simmons and S.G. Tilford

Vibrational Analysis - Spectrochim. Acta 19, 1473 (1963) - H. Perkampus and E. Baumgarten

Vibrational Analysis - J. Mol. Spectrosc. 14, 190 (1964) - J.D. Simmons, K.K. Innes and G.M. Begun

Phosphorescence - J. Mol. Spectrosc. 43, 239 (1972) - C.J. Marzzacco and E.F. Zalewski

Fluorescence - J. Mol. Spectrosc. 43, 239 (1972) - C.J. Marzzacco and E.F. Zalewski

Absorption Spectrum - J. Mol. Spectrosc. 43, 477 (1972) - K.K. Innes, A.H. Kalantar, A.Y. Khan and T.J. Durnick

UV Absorption in Vapour (High Resolution)- J. Mol. Spectrosc. 52, 130 (1974) - S.N. Thakur and K.K. Innes

Raman - J. Mol. Spectrosc. 60, 277 (1976)- K. Kamagawa and M. Ito

Fluorescence - J. Mol. Spectrosc. 60, 277 (1976) - K. Kamagawa and M. Ito

Absorption Spectrum - J. Mol. Spectrosc. 60, 277 (1976) - K. Kamagawa and M. Ito

Pyridazine $(C_4H_4N_2)$

UV Absorption in Vapour — Nature 187, 500 (1960) — K.K. Innes, J. Merritt, W. Tincher and S. Tilford

UV Absorption in Vapour — J. Mol. Spectrosc. 15, 407 (1965) — J.E. Parkin and K.K. Innes

UV Absorption in Vapour — J. Mol. Spectrosc. 24, 247 (1967) — K.K. Innes and R.M. Lucas, Jr

Vibrational Analysis — Spectrochim. Acta 23A, 2233 (1967) — H.D. Stidham and J.V. Tucci

UV Absorption in Vapour — J. Mol. Spectrosc. 22, 125 (1967) — K.K. Innes, J.P. Bryne and I.G. Ross

UV Absorption (Crystal) — J. Mol. Spectrosc. 31, 76 (1969) — K.K. Innes, H.D. McSwiney, J.D. Simmons and S.G. Tilford

UV Absorption in Vapour (High Resolution)— J. Mol. Spectrosc. 36, 114 (1970) — K.K. Innes, W.C. Tincher and E.F. Pearson

Vibrational Analysis (Crystal) — Spectrochim. Acta 29A, 781 (1973) — R.H. Larkin and H.D. Stidham

Pyridazine-3,6-D$_2$ $(C_4D_2H_2N_2)$

Vibrational Analysis — Spectrochim. Acta 23A, 2233 (1967) — H.D. Stidham and J.V. Tucci

Pyridazine-4,5-D$_2$ $(C_4D_2H_2N_2)$

Vibrational Analysis — Spectrochim. Acta 23A, 2233 (1967) — H.D. Stidham and J.V. Tucci

UV Absorption in Vapour — J. Mol. Spectrosc. 36, 114 (1970) — K.K. Innes, W.C. Tincher and E.F. Pearson

Pyridazine-D$_4$ $(C_4D_4N_2)$

Vibrational Analysis — Spectrochim. Acta 23A, 2233 (1967) — H.D. Stidham and J.V. Tucci

UV Absorption in Vapour — J. Mol. Spectrosc. 36, 114 (1970) — K.K. Innes, W.C. Tincher and E.F. Pearson

Pyridine (C_5H_5N)

UV Absorption in Vapour — J. Chem. Phys. 14, 101 (1946) — H. Sponer and J. Stucken

UV Absorption in Vapour — J. Chem. Phys. 17, 587 (1949) — J.H. Rush and H. Sponer

UV Absorption in Vapour — Disc. Faraday Soc. 9, 26 (1950) — E.F. Herrington

UV Absorption in Solution — Disc. Faraday Soc. 9, 14 (1950) — M. Kasha

UV Absorption in Vapour — J. Chem. Phys. 20, 1847 (1952) — J.H. Rush and H. Sponer

Vibrational Analysis — J. Chem. Phys. 21, 1170 (1953) — L. Corrsin, B.J. Lax and R.C. Lord

UV Absorption in Solution — J. Chem. Phys. 22, 1077 (1954) — H.P. Stephenson

UV Absorption in Different States — Indian J. Phys. 30, 480 (1956) — S.B. Banerjee

Raman — Indian J. Phys. 30, 519 (1956) — G.S. Kastha

Vibrational Analysis — Can. J. Chem. 35, 1183 (1957) — J.K. Wilmshurt and H.J. Bernstein

Raman — Proc. Indian Acad. Sci. 45A, 51 (1957) — P.G. Puranik and A.M. Jaya Rao

UV Absorption in Solution — Indian J. Phys. 32, 323 (1958) — S.B. Banerjee

Calculation — Opt. Spektrosk. 10, 457 (1961) — M.A. Kovner, Yu.S. Korosrelov and V.I. Bereszin

UV Absorption in Different States — Indian J. Phys. 35, 420 (1961) — T.N. Misra

Phosphorescence — Spectrochim. Acta 17, 30 (1961) — R. Shimada

Calculation — Trans. Faraday Soc. 59, 12 (1963) — D.A. Long, F.S. Murfin and E.L. Thomas

Vibrational Analysis — Spectrochim. Acta 19, 549 (1963) — J.H.S. Green, W. Kynaston and H.M. Paisley

Raman — J. Amer. Chem. Soc. 85, 3072 (1963) — A. Fratiello and J.P. Luongo

Calculation — Opt. Spektrosk. 15, 167 (1963) — V.I. Berezin

Calculation — J. Chem. Phys. 38, 127 (1963) — E.G. Zerbi, B. Crawford, Jr and J. Overend

Vibrational Analysis — Spectrochim. Acta 21, 747 (1965) — D.B. Cuntiffe-Jones

IR Spectrum — J. Mol. Spectrosc. 21, 217 (1966) — H. Takahashi, K. Mamola and E.K. Plyer

Vibrational Analysis — J. Mol. Struct. 1, 157 (1967) — J. Loisel and V. Lorenzelli

UV Absorption in Vapour — J. Mol. Spectrosc.
    <u>22</u>, 125 (1967) — K.K. Innes, J.P. Bryne
    and I.G. Ross
Vibrational Analysis — J. Chem. Phys. <u>51</u>,
    3762 (1969) — E. Castellucci, G. Sbrana
    and F.D. Verderame
Raman — J. Mol. Struct. <u>5</u>, 477 (1970) —
    N.T. McDevitt and W.G. Fateley
IR Spectrum — J. Mol. Spectrosc. <u>61</u>, 164
    (1976) — Y. Kakiuti, M. Akiyama, N.
    Saito and H. Saito
UV Absorption in Vapour — Spectrochim. Acta
    <u>34A</u>, 211 (1978) — Uriel Olsher

Pyridine-D$_2$ (C$_5$H$_3$D$_2$N)

IR Spectrum — J. Mol. Spectrosc. <u>61</u>, 164
    (1976) — Y. Kakiuti, M. Akiyama, N.
    Saito and H. Saito

Pyridine-D$_3$ (C$_5$H$_2$D$_3$N)

IR Spectrum — J. Mol. Spectrosc. <u>61</u>, 164
    (1976) — Y. Kakiuti, M. Akiyama, N.
    Saito and H. Saito

Pyridine-D$_4$ (C$_5$HD$_4$N)

Calculation — Trans. Faraday Soc. <u>59</u>, 783
    (1963) — D.A. Long and E.L. Thomas
IR Spectrum — J. Mol. Spectrosc. <u>61</u>, 164
    (1976) — Y. Kakiuti, M. Akiyama, N.
    Saito and H. Saito

Pyridine-2,4-D$_2$ (C$_5$H$_3$D$_2$N)

Calculation — Trans. Faraday Soc. <u>59</u>, 783
    (1963) — D.A. Long and E.L. Thomas

Pyridine-2,5-D$_2$ (C$_5$H$_3$D$_2$N)

Calculation — Trans. Faraday Soc. <u>59</u>, 783
    (1963) — D.A. Long and E.L. Thomas

Pyridine-D$_5$ (C$_5$D$_5$N)

Vibrational Analysis — Can. J. Chem. <u>35</u>,
    1183 (1957) — J.K. Wilmshurst and H.J.
    Bernstein
Calculation — Trans. Faraday Soc. <u>59</u>, 12
    (1963) — D.A. Long, F.S. Murfin and E.L.
    Thomas

IR Spectrum — J. Mol. Spectrosc. <u>61</u>, 164
    (1976) — Y. Kakiuti, M. Akiyama, N.
    Saito and H. Saito

2-Pyridinealdehyde   (C$_5$H$_4$NCHO)

UV Emission in Vapour — Curr. Sci. <u>43</u>, 745
    (1974) — M.R. Padhye and C.J. Jahagi-
    rdar
UV Absorption in Vapour — Curr. Sci. <u>44</u>,
    579 (1975) — M.R. Padhye and C.J.
    Jahagirdar

3-Pyridinealdehyde   (C$_5$H$_4$NCHO)

UV Emission in Vapour — Curr. Sci. <u>43</u>,
    745 (1974) — M.R. Padhye and C.J.
    Jahagirdar
UV Absorption in Vapour — Curr. Sci. <u>44</u>,
    579 (1975) — M.R. Padhye and C.J.
    Jahagirdar

4-Pyridinealdehyde   (C$_5$H$_4$NCHO)

UV Emission in Vapour — Curr. Sci. <u>43</u>,
    745 (1974) — M.R. Padhye and C.J.
    Jahagirdar
UV Absorption in Vapour — Curr. Sci. <u>44</u>,
    579 (1975) — M.R. Padhye and C.J.
    Jahagirdar

Pyrimidine   (C$_4$H$_4$N$_2$)

UV Absorption in Solution — J. Amer. Chem.
    Soc. <u>56</u>, 1728 (1934) — F.F. Heyroth
    and J.R. Loofbourow
UV Absorption in Vapour — J. Amer. Chem.
    Soc. <u>63</u>, 137 (1941) — F.M. Uber and
    R. Winters
UV Absorption in Solution — J. Amer. Chem.
    Soc. <u>63</u>, 137 (1941) — F.M. Uber and
    R. Winters
UV Absorption in Vapour — J. Chem. Phys.
    <u>9</u>, 777 (1941) — F.M. Uber
UV Absorption in Vapour — J. Chem. Phys.
    <u>17</u>, 1165 (1949) — F. Halverson and
    R.C. Hirt
IR Spectrum — J. Chem. Soc. 3062 (1950) —
    I.A. Brownlee
UV Absorption in Solution — J. Chem. Phys.
    <u>19</u>, 711 (1951) — F. Halverson and R.C.
    Hirt
Vibrational Analysis — J. Chem. Phys. <u>25</u>,
    597 (1956) — M. Ito, R. Shimada, T.
    Kuraishi and W. Mizushima

Vibrational Analysis — Spectrochim. Acta
9, 113 (1957) — R.C. Lord, A.L. Marston
and F.A. Miller
UV Absorption in Vapour — J. Chem. Soc.
1263 (1959) — S.F. Mason
UV Absorption in Vapour — Nature 187, 500
(1960) — K.K. Innes, J. Merritt, W.
Tincher and S. Tilford
Vibrational Analysis — Spectrochim. Acta
17, 30 (1961) — R. Shimada
UV Absorption in Vapour — J. Mol. Spectrosc.
13, 435 (1964) — J.D. Simmons and K.K.
Innes
Fluorescence — J. Chem. Phys. 43, 2902
(1965) — B.J. Cohen, H. Baba and L.
Goodman
UV Absorption in Vapour — J. Mol. Spectrosc.
15, 407 (1965) — J.E. Parkin and K.K.
Innes
Vibrational Analysis — Spectrochim. Acta
21, 747 (1965) — D.B. Cunliffe-Jones
Calculation — Opt. Spektrosk. 18, 32
(1965) — V.I. Berzin and S.K. Totapav
Vibrational Analysis — Spectrochim. Acta
22, 1831 (1966) — G. Sbrana, G. Adembri
and S. Califano
IR Spectrum — J. Mol. Spectrosc. 21, 217
(1966) — H. Takahashi, K. Mamola
and E.K. Plyer
Fluorescence — J. Chem. Phys. 46, 713
(1967) — B.J. Cohen and L. Goodman
Vibrational Analysis — J. Chim. Phys. 64,
1484 (1967) — R. Foglizzo and A. Novak
UV Absorption in Vapour — J. Mol. Spectrosc.
22, 125 (1967) — K.K. Innes, J.P.
Bryne and I.G. Ross
Vibrational Analysis — Spectrosc. Lett.
2, 165 (1969) — A. Foglizzo and A. Novak
UV Absorption in Vapour — J. Mol. Spectrosc.
31, 76 (1969) — K.K. Innes, H.D. McSwi-
ney, J.D. Simmons and S.G. Tilford
Fluorescence — Chem. Phys. Lett. 9, 514
(1971) — Y.H. Li and E.C. Lim
Phosphorescence — J. Mol. Spectrosc. 42,
75 (1972) — R.M. Hochstrasser and C.J.
Marzzacco
Fluorescence — J. Mol. Spectrosc. 42, 75
(1972) — R.M. Hochstrasser and C.J.
Marzzacco
Vibrational Analysis — Spectrochim. Acta
29A, 781 (1973) — R.H. Larkine and H.D.
Stidham

Pyrimidine-D$_4$ (C$_4$D$_4$N$_2$)

UV Absorption in Vapour — J. Mol. Spectrosc.
13, 435 (1964) — J.D. Simmons and K.K.
Innes

IR Spectrum — Spectrochim. Acta 22, 1831
(1966) — G. Sbrana, G. Adembri and S.
Califano
Raman — J. Mol. Spectrosc. 31, 76 (1969) —
K.K. Innes, H.D. McSwiney, J.D.
Simmons and S.G. Tilford
Phosphorescence — J. Mol. Spectrosc. 42,
75 (1972) — R.M. Hochstrasser and C.J.
Marzzacco
Fluorescence — J. Mol. Spectrosc. 42, 75
(1972) — R.M. Hochstrasser and C.J.
Marzzacco
Vibrational Analysis — Spectrochim. Acta
29A, 781 (1973) — R.H. Larkin and
H.D. Stidham

Pyrocatechol (C$_6$H$_4$(OH)$_2$)

See 1,2-Dihydroxybenzene

Pyrogallol (C$_6$H$_3$(OH)$_3$)

See 1,2,3-Trihydroxybenzene

Quinaldine (C$_9$H$_6$NCH$_3$)

See 2-Methylquinoline

Quinaxoline (C$_8$H$_6$N$_2$)

Vibrational Analysis — J. Amer. Chem. Soc.
74, 4834 (1952) — H. Culbentson, J.C.
Decius and B.E. Christesen
Vibrational Analysis — Spectrochim. Acta
20, 593 (1964) — W.L.F. Armarego,
A.R. Katritzky and B.J. Ridgewell
IR Spectrum — Spectrochim. Acta 22, 117
(1966) — W.L.F. Armarego, G.B. Barlin
and E. Spinner
UV Absorption in Vapour — Bull. Chem. Soc.
Jap. 41, 2608 (1968) — Y. Haregawa,
Y. Amako and H. Azumi
UV Absorption Spectrum — J. Mol. Spectrosc.
30, 149 (1969) — Y. Kaizu and M. Ito
UV Absorption Spectrum — J. Chem. Phys.
53, 3857 (1970) — R.W. Glass, L.C.
Robertson and J.A. Merritt
Vibrational Analysis — J. Mol. Spectroc.
36, 310 (1970) — R.W. Mitchell, R.W.
Glass and J.A. Merritt
Fluorescence — J. Chem. Phys. 61, 3895
(1974) — J.R. McDonald and L.E. Brus
Phosphorescence — J. Chem. Phys. 61, 3895
(1974) — J.R. McDonald and L.E. Brus

Quinoline (C_9H_7N)

Raman — Indian J. Phys. 10, 23 (1937) — K. Jatkar

Vibrational Analysis — J. Chem. Soc. 71, 443 (1950) — I. Ichishima

UV Absorption in Vapour — Bull. Chem. Soc. Jap. 29, 373 (1956) — N. Mataga, Y. Kaifu and M. Koizumi

Vibrational Analysis — J. Prakt. Chem. 5, 242 (1958) — H. Luther, D. Mootz and F. Radwitz

Calculation — Bull. Chem. Soc. Jap. 31, 459 (1958) — N. Mataga

Vibrational Analysis — Ann. Chim. (Rome) 49, 245 (1959) — P. Chirboli and A. Bertoluzza

UV Absorption in Solution — Z. Elektrochem. 65, 61 (1961) — H. Zimmerman and N. Joop

Emission Spectrum in Vapour — J. Mol. Spectrosc. 6, 305 (1961) — D.E. Freeman

Vibrational Analysis — Indian J. Phys. 35, 535 (1961) — K.K. Deb

UV Absorption in Solution — Spectrochim. Acta 18, 1441 (1962) — G. Coppens, C. Gillet, J. Nascilski and E.V. Donckt

Vibrational Analysis — Indian J. Phys. 36, 557 (1962) — K.K. Deb

UV Absorption in Vapour — J. Mol. Spectrosc. 9, 138 (1962) — J.M. Hollas

Vibrational Analysis — Indian J. Pure Appl. Phys. 7, 567 (1969) — R. Amni Amma, K.P.R. Nair and S. Nath Singh

UV Absorption in Vapour — Indian J. Pure Appl. Phys. 7, 567 (1969) — R. Amni Amma, K.P.R. Nair and S. Nath Singh

Vibrational Analysis — J. Mol. Spectrosc. 34, 56 (1970) — S.C. Wait, Jr and J.C. McNerney

Calculation — J. Mol. Spectrosc. 44, 118 (1972) — H. Baba and I. Yamazaki

UV Absorption in Vapour — J. Mol. Spectrosc. 47, 189 (1973) — U.T. Kreibich and Urs. P. Wild

Calculation — J. Mol. Spectrosc. 50, 457 (1974) — J.E. Ridley and M.C. Zerner

Quinoline-D_7 (C_9D_7N)

Absorption Spectrum — J. Mol. Spectrosc. 49, 201 (1974) — G. Fischer

Quinoxaline ($C_8H_6N_2$)

Vibrational Analysis — Z. Naturforsch. B15, 1 (1960) — H.H. Perkampus and A. Roders

IR Spectrum — Spectrochim. Acta 22, 117 (1966) — W.L.F. Armarego, G.B. Barlin and E. Spinner

UV Absorption in Vapour — Cull. Chem. Soc. Jap. 41, 2608 (1968) — Y. Hasegawa, Y. Amako and H. Azumi

UV Absorption Spectrum — J. Chem. Phys. 51, 5015 (1969) — R.H. Clarke, R.M. Hochstrasser and C.J. Marzzacco

UV Absorption in Vapour — J. Chem. Phys. 53, 3857 (1970) — R.W. Glass, L.C. Robertson and J.A. Merritt

Vibrational Analysis — J. Mol. Spectrosc. 36, 310 (1970) — R.W. Mitchell, R.W. Glass and J.A. Merritt

UV Absorption in Vapour — J. Mol. Spectrosc. 40, 397 (1971) — G. Fischer, A.D. Jordan and I.G. Ross

UV Absorption in Vapour — Aust. J. Chem. 24, 1107 (1971) — J.P. Bryne and I.G. Ross

Calculation — J. Mol. Spectrosc. 44, 118 (1972) — H. Baba and I. Yamazaki

Fluorescence — J. Chem. Phys. 58, 1247 (1973) — E.C. Lim and C-S. Swang

Fluorescence — Chem. Phys. Lett. 23, 87 (1973) — J.R. McDonald and L.E. Brus

UV Absorption Spectrum (Crystal) — J. Mol. Spectrosc. 45, 173 (1973) — A.D. Jordan, G. Fischer, K. Rokos and I.G. Ross

UV Absorption in Solid — J. Mol. Spectrosc. 46, 316 (1973) — A.D. Jordan and I.G. Ross

Calculation — J. Mol. Spectrosc. 50, 457 (1974) — J.E. Ridley and M.C. Zerner

Phosphorescence — J. Chem. Phys. 61, 3895 (1974) — J.R. McDonald and L.E. Brus

Fluorescence — J. Chem. Phys. 61, 3895 (1974) — J.R. McDonald and L.E. Brus

Raman — J. Mol. Spectrosc. 59, 396 (1976) - N. Ohta and M. Ito

Phosphorescence — J. Chem. Phys. 67, 7 (1977) — S. Yamandu and T. Azumi

Quinoxaline-D_6 ($C_8D_6N_2$)

UV Absorption in Vapour — J. Mol. Spectrosc. 40, 397 (1971) — G. Fischer, A.D. Jordan and I.G. Ross

UV Absorption Spectrum (Crystal) — J. Mol. Spectrosc. 45, 173 (1973) — A.D. Jordan, G. Fischer, K. Rokos and I.G. Ross

Resorcinol $(C_6H_4(OH)_2)$

See 1,3-Dihydroxybenzene

Salicylaldehyde (C_6H_4OHCHO)

See o-Hydroxybenzaldehyde

Styrene $(C_6H_5CH:CH_2)$

Fluorescence — J. Chem. Soc. 3315 (1923) — J.K. Marsh

Raman — C.R. Acad. Sci.(Paris) 194, 1736 (1932) — N. Bourguel

IR Spectrum — J. Res. Natl. Bur. Stand. (U.S.) A15, 295 (1935) — R. Sair and W.W. Cobletz

IR Spectrum — Physics (N.Y.) 7, 399 (1936)— D. Williams

IR Spectrum — J. Amer. Chem. Soc. 65, 803 (1943) — K.S. Pitzer, L. Guttman and E.F. Westrum

Vibrational Analysis — J. Amer. Chem. Soc. 68, 2209 (1946) — K.S. Pitzer, L. Guttman and E.F. Westrum

UV Absorption in Solution — J. Chem. Phys. 18, 1168 (1950) — J.R. Platt

UV Absorption in Vapour — J. Amer. Chem. Soc. 72, 5260 (1950) — W.W. Robertson, J.F. Maesic and F.A. Matsen

UV Absorption in Vapour — J. Amer. Chem. Soc. 75, 5055 (1953) — J.V. Morgan

Vibrational Analysis — J. Chem. Phys. 22, 236 (1954) — S. Nagakura and J. Tanaka

Raman — Indian J. Phys. 28, 365 (1954) — N.K. Roy

UV Absorption in Different States — Indian J. Phys. 30, 321 (1956) — S.K. Sen

UV Emission in Vapour — Indian J. Pure Appl. Phys. 4, 1 (1966) — R.N. Singh

Vibrational Analysis — Spectrochim. Acta 23A, 895 (1967) — F.A. Miller, W.G. Fateley and R.E. Wilkowski

Vibrational Analysis — Appl. Spectrosc. 22, 650 (1968) — W.G. Fateley, G.L. Carlson and F.E. Dickson

UV Absorption in Vapour — J. Mol. Spectrosc. 35, 413 (1970) — A.R. Hartford and J.R. Lombardi

Raman — J. Mol. Struct. 5, 477 (1970) — N.T. McDevitt and W.G. Fateley

Vibrational Analysis — Spectrochim. Acta 26A, 1097 (1970) — W.D. Mross and G. Zundel

Vibrational Analysis — J. Mol. Struct. 30, 1 (1976) — W.M. Rolowski, P.J. Mjoberg and S.O. Ljunggren

Vibrational Analysis — J. Mol. Spectrosc. 63, 466 (1976) — D.A. Condirston and J.D. Laposa

Calculation — Indian J. Phys. 51B, 178 (1977) — B.J. Ansari

Styrene-D₁ $(C_6H_4DCH:CH_2)$

Vibrational Analysis — Spectrochim. Acta 26A, 1109 (1970) — W.D. Mross and G. Zundel

Vibrational Analysis — Spectrochim. Acta 33A, 249 (1977) — J.H.S. Green and D.J. Harrison

Styrene-D₃ $(C_6H_2D_3CH:CH_2)$

Vibrational Analysis — J. Mol. Spectrosc. 63, 466 (1976) — D.A. Condirston and J.D. Laposa

Styrene-D₅ $(C_6D_5CH:CH_2)$

Vibrational Analysis — J. Mol. Spectrosc. 63, 466 (1976) — D.A. Condirston and J.D. Laposa

Styrene-D₈ $(C_6D_5CH:CD_2)$

Vibrational Analysis — Spectrochim. Acta 26A, 1109 (1970) — W.D. Mross and G. Zundel

Vibrational Analysis — J. Mol. Spectrosc. 63, 466 (1976) — D.A. Condirston and J.D. Laposa

Vibrational Analysis — Spectrochim. Acta 33A, 249 (1977) — J.H.S. Green and D.J. Harrison

N-Sulfinylaniline (C_6H_5NSO)

Vibrational Analysis — Spectrochim. Acta 17, 933 (1961) — C.V. Stephenson, W.C. Coburn and W.S. Wilcox

Sym-Tetrazine $(C_2H_2N_4)$

See 1,2,4,5-Tetrazine

Sym-Triazine ($C_3H_3N_3$)

See   1,3,5-Triazine

Sym-Tribromobenzene ($C_6H_3Br_3$)

See   1,3,5-Tribromobenzene

Sym-Trichlorobenzene ($C_6H_3Cl_3$)

See   1,3,5-Trichlorobenzene

Sym-Trifluorobenzene ($C_6H_3F_3$)

See   1,3,5-Trifluorobenzene

1,2,3,4-Tetrachlorobenzene ($C_6H_2Cl_4$)

Vibrational Analysis — Spectrochim. Acta
   19, 1739 (1963) — J.R. Scherer and
   J.C. Evans

1,2,3,5-Tetrachlorobenzene ($C_6H_2Cl_4$)

Vibrational Analysis — Spectrochim. Acta
   19, 1739 (1963) — J.R. Scherer and J.C.
   Evans

1,2,4,5-Tetrachlorobenzene ($C_6H_2Cl_4$)

Vibrational Analysis — Spectrochim. Acta
   19, 1739 (1963) — J.R. Scherer and
   J.C. Evans

1,2,4,5-Tetrachlorobenzene-D$_1$
(C_6HDCl_4)

Vibrational Analysis — Spectrochim. Acta
   19, 1739 (1963) — J.R. Scherer and
   J.C. Evans

1,2,4,5-Tetrachlorobenzene-D$_2$
($C_6D_2Cl_4$)

Vibrational Analysis — Spectrochim. Acta
   19, 1739 (1963) — J.R. Scherer and
   J.C. Evans

1,2,3,4-Tetrafluorobenzene ($C_6H_2F_4$)

UV Absorption in Vapour — J. Chem. Phys.
   21, 1457 (1953) — E.E. Ferguson, R.L.
   Hudson, J.R. Nielsen and D.C. Smith
Vibrational Analysis — Spectrochim. Acta
   18, 915 (1962) — D. Steele
Fluorescence — Can. J. Chem. 48, 2324
   (1970) — B.H. Scholz and I. Unger
Calculation — J. Mol. Spectrosc. 48, 446
   (1973) — V.J. Eaton and D. Steele
Vibrational Analysis — Spectrochim. Acta
   32A, 1185 (1976) — J.H.S. Green and
   D.J. Harrison
Calculation — J. Mol. Struct. 32, 93
   (1976) — F. Torok, A. Hegedus, K. Kosa
   and P. Puley

1,2,3,5-Tetrafluorobenzene ($C_6H_2F_4$)

Vibrational Analysis — Spectrochim. Acta
   18, 915 (1962) — D. Steele
Fluorescence — Can. J. Chem. 48, 2324
   (1970) — B.H. Scholz and I. Unger
Calculation — J. Mol. Spectrosc. 48, 446
   (1973) — V.J. Eaton and D. Steele
Calculation — J. Mol. Struct. 32, 93
   (1976) — F. Torok, A. Hegedus, K. Kosa
   and P. Puley
Vibrational Analysis — Spectrochim. Acta
   32A, 1185 (1976) — J.H.S. Green and
   D.J. Harrison

1,2,4,5-Tetrafluorobenzene ($C_6H_2F_4$)

Vibrational Analysis — J. Chem. Phys. 21,
   1464 (1953) — E.E. Ferguson, R.L.
   Hudson, J.R. Nielsen and D.C. Smith
Vibrational Analysis — Trans. Faraday Soc.
   55, 369 (1959) — D. Steele and D.H.
   Whiffen
Vibrational Analysis — Spectrochim. Acta
   16, 368 (1960) — D. Steele and D.H.
   Whiffen
Fluorescence — Can. J. Chem. 48, 2324
   (1970) — B.H. Scholz and I. Unger
Calculation — J. Mol. Spectrosc. 48, 446
   (1973) — V.J. Eaton and D. Steele
Calculation — J. Mol. Struct. 32, 93
   (1976) — F. Torok, A. Hegedus, K. Kosa
   and P. Puley
Vibrational Analysis — Spectrochim. Acta
   32A, 1185 (1976) — J.H.S. Green and
   D.J. Harrison

2,3,5,6-Tetrafluorotoluene ($C_6HF_4CH_3$)

Vibrational Analysis — Spectrochim. Acta
25A, 1563 (1969) — R.T. Bailey

1,2,3,4-Tetramethylbenzene ($C_6H_2(CH_3)_4$)

Vibrational Analysis — Angew. Chem. A59,
142 (1947) — H. Fromherz and H. Bueren
Vibrational Analysis — Spectrochim. Acta
4, 373 (1951) — C.C. Cannon and G.B.
B.M. Sutherland
IR Spectrum — J. Phys. Chem. 61, 730
(1957) — S.H. Hestings and D.E.
Nicolson
Fluorescence — Chem. Phys. Lett. 14, 404
(1972) — G.M. Breuer and E.K.C. Lee
Calculation — J. Mol. Spectrosc. 48, 446
(1973) — V.J. Eaton and D. Steele

1,2,3,5-Tetramethylbenzene ($C_6H_2(CH_3)_4$)

IR Spectrum — J. Phys. Chem. 61, 730
(1957) — S.H. Hestings and D.E.
Nicolson
Vibrational Analysis — Izv. Akad. Nauk.
1444 (1961) — Kh.E. Sterin, V.T. Aleksanian, S.A. Ukholin, O.V. Bragin, A.E.
Gavrilova, S.V. Zotova, A.L. Liebermann, E.A. Mikhailova, E.N. Smirnova,
O.D. Sterhigov and B.A. Kozanski
Fluorescence — Chem. Phys. Lett. 14, 404
(1972) — G.M. Breuer and E.K.C. Lee

1,2,4,5-Tetramethylbenzene ($C_6H_2(CH_3)_4$)

Vibrational Analysis — Opt. Spektrosk. 1,
348 (1956) — M.A. Kovner
IR Spectrum — J. Phys. Chem. 61, 730
(1957) — S.H. Hestings and D.E.
Nicolson

1,2,4,5-Tetrazine ($C_2H_2N_4$)

UV Absorption Spectrum — J. Chem. Soc. 1240
(1959) — S.F. Mason
UV Absorption Spectrum — J. Chem. Soc. 1263
(1959) — S.F. Mason
UV Absorption Spectrum — J. Chem. Soc. 1269
(1959) — S.F. Mason
Calculation — Spectrochim. Acta 16, 900
(1960) — B. Crawford and S. Califano
Vibrational Analysis — Spectrochim. Acta
17, 155 (1961) — J.E. Lancaster,
R.F. Stamm and N.B. Colthup

IR Spectrum — J. Chem. Phys. 35, 1939
(1961) — G.H. Spencer, P.C. Cross and
K.B. Wiberg
Fluorescence — J. Chem. Phys. 35, 1925
(1961) — G.H. Spencer, P.C. Cross and
K.B. Wiberg
Fluoresnce — J. Chem. Phys. 36, 548
(1962) — M. Chaudhury and L. Goodman
Fluorescence — J. Chem. Phys. 38, 2979
(1963) — M. Chaudhury and L. Goodman
UV Absorption in Vapour — J. Mol. Spectrosc.
22, 125 (1967) — K.K. Innes, J.P.
Bryne and I.G. Ross
Vibrational Analysis — J. Mol. Spectrosc.
26, 458 (1968) — L.A. Frank, A.J.
Merer and K.K. Innes
UV Absorption in Vapour — Proc. Roy. Soc.
A302, 271 (1968) — A.J. Merer and
K.K. Innes
UV Absorption Spectrum — Can. J. Phys.
47, 234 (1969) — J.M. Brown
Vibrational Analysis — Spectrochim. Acta
27A, 747 (1971) — W.D. Sigworth and
E.L. Pace
UV Absorption in Vapour — J. Mol. Spectrosc.
39, 115 (1971) — D.T. Livak and K.K.
Innes
UV Absorption in Vapour (High Resolution) —
J. Mol. Spectrosc. 40, 177 (1971) —
K.K. Innes, A.Y. Khan and D.T. Livak
UV Absorption in Vapour — J. Mol. Spectrosc.
43, 477 (1972) — K.K. Innes, A.H.
Kahartar, A.Y. Khan and T.J. Durnick
Fluorescence — Chem. Phys. Lett. 28, 364
(1974) — J.M. Meyling, R.P. Vander
Wert and D.A. Wierzma
Calculation — Spectrochim. Acta 31A, 979
(1975) — Y. Kawaguchi and R.H. Mann

1,2,4,5-Tetrazine-D_2 ($C_2D_2N_4$)

Vibrational Analysis — J. Mol. Spectrosc.
26, 458 (1968) — L.A. Franks, A.J.
Merer and K.K. Innes
Calculation — Spectrochim. Acta 31A, 979
(1975) — Y. Kawaguchi and R.H. Mann

o-Thiocresol ($C_6H_4CH_3SH$)

IR Spectrum — Indian J. Phys. 33, 295
(1959) — R.N. Bapat
Raman — Indian J. Phys. 33, 329 (1959) —
R.N. Bapat

m-Thiocresol ($C_6H_4CH_3SH$)

IR Spectrum — Indian J. Phys. $\underline{33}$, 295 (1959) — R.N. Bapat
Raman — Indian J. Phys. $\underline{33}$, 329 (1959) — R.N. Bapat

p-Thiocresol ($C_6H_4CH_3SH$)

Vibrational Analysis — Spectrochim. Acta $\underline{12}$, 305 (1958) — C. Garrigou-Lagrange, J.M. Lebas and M.L. Josien
Vibrational Analysis — Spectrochim. Acta $\underline{13}$, 225 (1959) — J.M. Lebas, C. Garrigou-Lagrange and M.L. Josien
IR Spectrum — Indian J. Phys. $\underline{33}$, 295 (1959) — R.N. Bapat
Raman — Indian J. Phys. $\underline{33}$, 329 (1959) — R.N. Bapat
Vibrational Analysis — Spectrochim. Acta $\underline{17}$, 503 (1961) — R.A. Nyquist and J.C. Evans
Vibrational Analysis — Spectrochim. Acta $\underline{26A}$, 1515 (1970) — J.H.S. Green, D.J. Harrison and W. Kynaston
UV Absorption in Vapour — Indian J. Phys. $\underline{49}$, 703 (1975) — P.K. Mallik

Thiophenol (C_6H_5SH)

See   Benzenethiol

o-Tolualdehyde ($C_6H_4CH_3CHO$)

UV Absorption in Solution — J. Chem. Soc. 1088 (1913) — J.E. Purvis
UV Absorption in Different States — J. Chem. Soc. 2282 (1914) — J.E. Purvis
UV Emission in Vapour — J. Chem. Soc. 2401 (1929) — R. Russel, J.C. MacMaster and A.W. Stewart
Raman — Z. Phys. Chem. $\underline{38B}$, 119 (1937) — L. Kahovec and K.W.F. Kohlrausch
UV Absorption in Vapour — J. Mol. Spectrosc. $\underline{20}$, 282 (1966) — V.B. Singh and I.S. Singh
UV Absorption in Vapour — Curr. Sci. $\underline{36}$, 603 (1967) — V.B. Singh and I.S. Singh
Vibrational Analysis — Indian J. Pure Appl. Phys. $\underline{6}$, 682 (1968) — V.B. Singh and I.S. Singh
Vibrational Analysis — Indian J. Phys. $\underline{42}$, 266 (1968) — V.B. Singh and I.S. Singh
UV Emission in Vapour — Curr. Sci. $\underline{38}$, 491 (1969) — K. Singh and V.B. Singh

m-Tolualdehyde ($C_6H_4CH_3CHO$)

UV Absorption in Solution — J. Chem. Soc. 1088 (1913) — J.E. Purvis
UV Absorption in Different States — J. Chem. Soc. 2282 (1914) — J.E. Purvis
UV Emission in Vapour — J. Chem. Soc. 2041 (1929) — R. Russel, J.C. MacMaster and A.W. Stewart
Raman — Z. Phys. Chem. $\underline{38B}$, 119 (1937) — L. Kahovec and K.W.F. Kohlrausch
UV Absorption in Vapour — J. Mol. Spectrosc. $\underline{20}$, 282 (1966) — V.B. Singh and I.S. Singh
UV Absorption in Vapour — Curr. Sci. $\underline{36}$, 603 (1967) — V.B. Singh and I.S. Singh
Vibrational Analysis — Indian J. Phys. $\underline{42}$, 266 (1968) — V.B. Singh and I.S. Singh
Vibrational Analysis — Indian J. Pure Appl. Phys. $\underline{6}$, 682 (1968) — V.B. Singh and I.S. Singh
UV Emission in Vapour — Curr. Sci. $\underline{38}$, 491 (1969) — K. Singh and V.B. Singh

p-Tolualdehyde ($C_6H_4CH_3CHO$)

UV Absorption in Solution — J. Chem. Soc. 1088 (1913) — J.E. Purvis
UV Absorption in Different States — J. Chem. Soc. 2282 (1914) — J.E. Purvis
UV Emission in Vapour — J. Chem. Soc. 2401 (1929) — R. Russel, J.C. MacMaster and A.W. Stewart
Raman — Z. Phys. Chem. $\underline{38B}$, 119 (1937) — L. Kahovec and K.W.F. Kohlrausch
UV Absorption in Vapour — J. Sci. Res. BHU (India) $\underline{14}$, 32 (1963) — N.L. Singh and R.N. Singh
UV Absorption in Vapour — J. Mol. Spectrosc. $\underline{20}$, 282 (1966) — V.B. Singh and I.S. Singh
UV Absorption in Vapour — Curr. Sci. $\underline{36}$, 603 (1967) — V.B. Singh and I.S. Singh
Vibrational Analysis — Indian J. Phys. $\underline{42}$, 266 (1968) — V.B. Singh and I.S. Singh
Vibrational Analysis — Indian J. Pure Appl. Phys. $\underline{6}$, 682 (1968) — V.B. Singh and I.S. Singh
UV Emission in Vapour — Curr. Sci $\underline{38}$, 49 (1969) — K. Singh and V.B. Singh

Toluene ($C_6H_5CH_3$)

UV Absorption in Solution — Ann. Chim. (Paris) 11, 287 (1927) — J. Savart
UV Absorption in Vapour — Bull. Chem. Soc. Jap. 11, 346 (1936) — K. Masaki
UV Absorption in Vapour — J. Chem. Phys. 10, 672 (1942) — H. Sponer
IR Spectrum — J. Amer. Chem. Soc. 65, 803 (1943) — K.S. Pitzer
UV Absorption in Vapour — J. Chem. Phys. 14, 511 (1946) — N. Ginsberg, F.A. Matsen and W.W. Robertson
UV Absorption in Vapour — J. Amer. Chem. Soc. 70, 577 (1948) — A.R. Choppin and C.H. Smith
UV Emission in Vapour — Indian J. Phys. 23, 339 (1949) — M.R. Padhye and R.K. Asundi
UV Absorption in Vapour — Indian J. Phys. 23, 331 (1949) — M.R. Padhye
UV Absorption in Vapour — Sci. Cult. (Calcutta) 14, 291 (1949) — R.K. Asundi and M.R. Padhye
Fluorescence — J. Chem. Phys. 18, 1403 (1950) — A.M. Bass
Raman — Indian J. Phys. 24, 111 (1950) — A.K. Ray
UV Absorption in Different States — Indian J. Phys. 25, 261 (1951) — H.N. Swamy
UV Emission in Vapour — J. Sci. Ind. Res. (India) 11B, 200 (1952) — S.N. Garg
UV Emission in Vapour — J. Sci. Ind. Res. (India) 11B, 447 (1952) — S.N. Garg
UV Absorption in Vapour — J. Chem. Phys. 20, 1248 (1952) — C.D. Cooper and H. Sponer
Raman (Solid) — Indian J. Phys. 30, 313 (1956) — G.S. Kastha
Vibrational Analysis — Can. J. Chem. 35, 911 (1957) — J.K. Wilmshurt and H.J. Bernstein
Vibrational Analysis — Opt. Spektrosk. 5, 134 (1958) — M.A. Kovner and G.V. Peregudov
Vibrational Analysis — Spectrochim. Acta 16, 106 (1960) — N. Fuson, C. Garrigou-Lagrange and M.L. Josien
Vibrational Analysis — Opt. Spektrosk. 9, 311 (1960) — A.M. Bogomolov
IR Spectrum — J. Chem. Soc. 2236 (1961) — J.H.S. Green
Fluorescence — Proc. Roy. Soc. A283, 83 (1965) — J.B. Birks, C.L. Braga and M.D. Lumb
IR Spectrum — Spectrochim. Acta 22, 501 (1966) — W.R. McWhinnie and R.C. Poller
Fluorescence — J. Chem. Phys. 49, 1705 (1968) — C.S. Burton and W.A. Noyes

Fluorescence — J. Chem. Phys. 51, 3130 (1969) — G.M. Brewer and E.K.C. Lee
Vibrational Analysis — Appl. Spectrosc. 23, 242 (1969) — M.C. Goldberg. R.L. Wershaw, H. Babad and H.W. Mueller
Raman — J. Mol. Struct. 5, 477 (1970) — N.T. McDevitt and W.G. Fateley
Fluorescence — Chem. Phys. Lett. 6, 352 (1970) — C.S. Burton and H.E. Hunziker
Vibrational Analysis — Spectrochim. Acta 27A, 2073 (1971) — C. La Lau and R.G. Snyder
Fluorescence — Ber. Bunsenges. Phys. Chem. 75, 450 (1971) — J.M. Blondeau and M. Stockburger
Emission Spectrum (Electron Impact) — Chem. Phys. 233 (1972) — T. Ogawa, M. Tsuji, M. Toyoda and N. Ishibashi
Fluorescence — J. Chem. Phys. 56, 1011 (1972) — L.J. Leyshon and A. Reiser
Fluorescence (Electron Impact) — Bull. Chem. Soc. Jap. 46, 2637 (1973) — T. Ogawa, M. Tsuji, M. Toyoda and N. Ishibashi
Calculation — J. Mol. Struct. 16, 365 (1973) — R.T.C. Brownlee, D.G. Cameron, R.D. Topsom, A.R. Katritzky and A.J. Sparrow
Fluorescence — J. Chem. Phys. 61, 1782 (1974) — K.C. Smith, J.A. Schiavone and R.S. Freund
Fluorescence (Electron Impact) — J. Chem. Phys. 61, 1789 (1974) — K.C. Smith, J.A. Schiavone and R.S. Freund
Fluorescence — J. Chem. Phys. 61, 2160 (1974) — K.C. Smith, J.A. Schiavone and R.S. Freund

Toluene-α-D_1 ($C_6H_5CH_2D$)

UV Absorption in Vapour — J. Amer. Chem. Soc. 70, 577 (1948) — A.R. Choppin and C.H. Smith
Vibrational Analysis — J. Amer. Chem. Soc. 71, 4045 (1949) — J. Turkevich, H.A. McKenzie, L. Friedman and R. Spurr
Vibrational Analysis — J. Amer. Chem. Soc. 72, 3260 (1950) — C.H. Smith, A.R. Choppin and D.A. Nance

Toluene-α, α, α-D_3 ($C_6H_5CD_3$)

Vibrational Analysis — Can. J. Chem. 35, 911 (1957) — J.K. Wilmshurt and H.J. Bernstein
Vibrational Analysis — J. Chem. Phys. 27, 740 (1957) — N.A. Narasimhan, J.R. Nielsen and R. Theimer

116

Vibrational Analysis — Spectrochim. Acta
16, 106 (1960) — N. Fuson, C. Garrigou-
Lagrange and M.L. Josien
Vibrational Analysis — Spectrochim. Acta
27A, 2073 (1971) — C. La Lau and R.G.
Snyder
IR Spectrum — Spectrochim. Acta 31A, 245
(1975) — A.B. Dempster, D.B. Powell
and N. Sheppard
UV Absorption in Vapour — Indian J. Pure
Appl. Phys. 14, 81 (1976) — G.N.R.
Tripathi, B.N. Tiwari and R.M. Verma

Toluene-o-D₁ (C₆H₄DCH₃)

UV Absorption in Vapour — J. Amer. Chem.
Soc. 70, 577 (1948) — A.R. Choppin
and C.H. Smith
Vibrational Analysis — J. Amer. Chem. Soc.
71, 4045 (1949) — J. Turkevich, H.A.
McKenzie, L. Friedman and R. Spurr
Vibrational Analysis — J. Amer. Chem. Soc.
72, 3260 (1950) — C.H. Smith, A.R.
Choppin and D.A. Nance
Vibrational Analysis — J. Chim. Phys. 64,
1473 (1967) — G. Lucazeau and J.M.
Lebas
Vibrational Analysis — Appl. Spectrosc.
23, 242 (1969) — M.C. Goldberg, R.L.
Wershaw, H. Babad and H.W. Mueller

Toluene-m-D₁ (C₆H₄DCH₃)

UV Absorption in Vapour — J. Amer. Chem.
Soc. 70, 577 (1948) — A.R. Choppin
and C.H. Smith
Vibrational Analysis — J. Amer. Chem. Soc.
71, 4045 (1949) — J. Turkevich, H.A.
McKenzie, L. Friedman and R. Spurr
Vibrational Analysis — J. Amer. Chem. Soc.
72, 3260 (1950) — C.H. Smith, A.R.
Choppin and D.A. Nance
Vibrational Analysis — J. Chim. Phys. 64,
1473 (1967) — G. Lucazeau and J.M.
Lebas
Vibrational Analysis — Appl. Spectrosc. 23,
242 (1969) — M.C. Goldberg, R.L.
Wershaw, H. Babad and H.W. Mueller

Toluene-p-D₁ (C₆H₄DCH₃)

UV Absorption in Vapour — J. Amer. Chem.
Soc. 70, 577 (1948) — A.R. Choppin
and C.H. Smith

Vibrational Analysis — J. Amer. Chem. Soc.
71, 4045 (1949) — J. Turkevich, H.A.
McKenzie, L. Friedman and R. Spurr
Vibrational Analysis — J. Amer. Chem. Soc.
72, 3260 (1950) — C.H. Smith, A.R.
Choppin and D.A. Nance
Vibrational Analysis — J. Chim. Phys. 64,
1473 (1967) — G. Lucazeau and J.M.
Lebas

Toluene-D₅ (C₆D₅CH₃)

Vibrational Analysis — Spectrochim. Acta
16, 106 (1960) — N. Fuson, C. Garrigou-
Lagrange and M.L. Josien
IR Spectrum — Spectrochim. Acta 28A, 373
(1972) — A.B. Dempster, D.B. Powell
and N. Sheppard
Fluorescence — J. Chem. Phys. 59, 4453
(1973) — D.M. Haaland and G.C. Nieman
Vibrational Analysis — J. Mol. Spectrosc.
54, 223 (1975) — A.P. Hitchcock and
J.D. Laposa
IR Spectrum — Spectrochim. Acta 31A, 245
(1975) — A.B. Dempster, D.B. Powell
and N. Sheppard

Toluene-D₈ (C₆D₅CD₃)

Vibrational Analysis — Opt. Spektrosk. 5,
134 (1958) — M.A. Kovner and G.V.
Peregudov
Vibrational Analysis — Spectrochim. Acta
16, 106 (1960) — N. Fuson, C. Garrigou-
Lagrange and M.L. Josien
Vibrational Analysis — Spectrochim. Acta
27A, 2073 (1971) — C. La Lau and R.G.
Snyder
UV Absorption in Vapour — Indian J. Pure
Appl. Phys. 14, 81 (1976) — G.N.R.
Tripathi, B.N. Tiwari and R.M. Verma

Toluenethiol (C₆H₄CH₃SH)

See   Thiocresol

Toluidine (C₆H₄CH₃NH₂)

See   Methylaniline

o-Tolunitrile (C₆H₄CNCH₃)

UV Absorption in Solution — J. Chem. Soc.
501 (1915) — J.E. Purvis

UV Absorption in Different States — Indian
J. Phys. $\underline{29}$, 561 (1955) — S.K. Sen
Phosphorescence — Spectrochim. Acta $\underline{18}$,
1201 (1962) — K. Takai and Y. Kanda
UV Absorption in Vapour — J. Sci. Ind.
Res. (India) — $\underline{12B}$, 241 (1962) — M.R.
Padhye and T.S. Varadarajan
UV Absorption in Vapour — J. Sci. Res. BHU
(India) $\underline{14}$, 32 (1963) — N.L. Singh
and R.N. Singh
UV Emission in Vapour — J. Sci. Res.
BHU (India) $\underline{14}$, 32 (1963) — N.L.
Singh and R.N. Sirgh

m-Tolunitrile ($C_6H_4CNCH_3$)

Phosphorescence — Spectrochim. Acta $\underline{18}$,
1201 (1962) — K. Takai and Y. Kanda
UV Absorption in Vapour — J. Sci. Ind.
Res. (India) $\underline{21B}$, 241 (1962) — M.R.
Padhye and T.S. Varadarajan
UV Absorption in Vapour — J. Sci. Res.
B.H.U. (India) $\underline{14}$, 45 (1963) — R.N.
Singh
UV Emission in Vapour — Indian J. Pure
Appl. Phys. $\underline{3}$, 101 (1965) — R.N. Singh
and N.L. Singh

p-Tolunitrile ($C_6H_4CNCH_3$)

UV Absorption in Solution — J. Chem. Soc.
501 (1915) — J.E. Purvis
Raman — Monatsh. Chem. $\underline{63}$, 427 (1933) —
K.W.F. Kohlrausch and A. Pongratz
IR Spectrum — J. Phys. Radium $\underline{8}$, 487
(1938) — J. Lecomte
IR Spectrum — J. Phys. Radium $\underline{9}$, 13
(1939) — J. Lecomte
UV Absorption in Different States — Indian
J. Phys. $\underline{29}$, 561 (1955) — S.K. Sen
Phosphorescence — Spectrochim. Acta $\underline{18}$,
1201 (1962) — K. Takai and Y. Kanda
UV Absorption in Vapour — J. Sci. Ind.
Res. (India) — $\underline{21B}$, 241 (1962) — M.R.
Padhye and T.S. Varadarajan
UV Emission in Vapour — Indian J. Pure
Appl. Phys. $\underline{5}$, 281 (1967) — R.N. Singh

p-Toluoylfluoride ($C_6H_4CH_3COF$)

Vibrational Analysis — Acta Phys. Aust.
$\underline{1}$, 352 (1948) — H. Seewann-Albert and
L. Kahovec

Toluquinone ($C_6H_3CH_3O_2$)

Vibrational Analysis — Spectrochim. Acta
$\underline{32A}$, 1235 (1976) — M.F. Merienne-
Lafore
Visible Absorption (Crystal) — Spectrochim.
Acta $\underline{33A}$, 453 (1977) — M.F. Merienne-
Lafore

Toluquinone-D_3 ($C_6D_3CH_3O_2$)

Vibrational Analysis — Spectrochim. Acta
$\underline{32A}$, 1235 (1976) — M.F. Merienne-
Lafore
Visible Absorption (Crystal) — Spectrochim.
Acta $\underline{33A}$, 453 (1977) — M.F. Merienne-
Lafore

Toluquinone-D_6 ($C_6D_3CD_3O_2$)

Vibrational Analysis — Spectrochim. Acta
$\underline{32A}$, 1235 (1976) — M.F. Merienne-
Lafore
Visible Absorption (Crystal) — Spectrochim.
Acta $\underline{33A}$, 453 (1977) — M.F. Merienne-
Lafore

1,3,5-Triazine ($C_3H_3N_3$)

Vibrational Analysis — J. Phys. Chem. $\underline{58}$,
1078 (1954) — J. Goubeau, E.L. Jahn,
A. Kreutzberger and C. Grundmann
Vibrational Analysis — J. Chem. Phys. $\underline{22}$,
1149 (1954) — J.E. Lancaster and N.B.
Colthup
Vibrational Analysis — J. Chem. Phys. $\underline{22}$,
1280 (1954) — R.F. Stamm and J.E.
Lancaster
Vibrational Analysis — J. Phys. Chem. $\underline{58}$,
1078 (1954) — J. Goubeau, E.L. Jahn,
A. Kreutzberger and C. Grundmann
UV Absorption in Vapour — J. Chem. Phys.
$\underline{22}$, 1148 (1954) — R.C. Hirt, F.
Halverson and R.G. Schmitt
Raman — Can. J. Phys. $\underline{34}$, 1016 (1956) —
J.E. Lancaster and B.P. Stoicheff
Calculation — Spectrochim. Acta $\underline{16}$, 900
(1960) — B.L. Crawford and S. Califano
Vibrational Analysis — Spectrochim. Acta
$\underline{17}$, 155 (1961) — J.E. Lancaster, R.F.
Stamm and N.B. Colthup
UV Absorption in Vapour — J. Chem. Phys.
$\underline{35}$, 1219 (1961) — J.S. Brinen and L.
Goodman
UV Absorption in Vapour — Spectrochim. Acta
$\underline{17}$, 863 (1962) — J.S. Brinen, R.C.
Hirt and R.G. Schmitt

UV Absorption in Vapour — J. Mol. Spectrosc.
22, 125 (1967) — K.K. Innes, J.P. Bryne
and I.G. Ross
Phosphorescence — J. Mol. Spectrosc. 26,
24 (1968) — J.B. Gollivan, J.S. Brinen
and J.G. Koren
UV Absorption in Vapour (High Resolution)—
J. Mol. Spectrosc. 39, 400 (1971) —
Y. Udagawa, M. Ito and S. Nakakura
IR Spectrum — J. Mol. Spectrosc. 62, 373
(1976) — S.J. Daunt and H.F. Shurvell

1,3,5-Triazine-D_3 ($C_3D_3N_3$)

Vibrational Analysis — Spectrochim. Acta
17, 155 (1961) — J.E. Lancaster, R.F.
Stamm and N.B. Colthup
UV Absorption in Vapour — Spectrochim.
Acta 17, 863 (1962) — J.S. Brinen,
R.C. Hirt and R.G. Schmitt
IR Spectrum — J. Mol. Spectrosc. 62, 373
(1976) — S.J. Daunt and H.F. Shurvell

1,3,5-Tribromobenzene ($C_6H_3Br_3$)

Vibrational Analysis — Spectrochim. Acta
18, 1579 (1962) — J.R. Scherer, J.C.
Evans and W.W. Muelder
IR Spectrum — Spectrochim. Acta 19, 844
(1963) — E.F. Mooney
Vibrational Analysis (Crystal) — Spectro-
chim. Acta 27A, 405 (1971) — D.E.
Muller, J. Inoue, R.H. Larkin and
H.D. Stidham

1,3,5-Tribromobenzene-D_1 ($C_6H_2DBr_3$)

Vibrational Analysis — Spectrochim. Acta
18, 1579 (1962) — J.R. Scherer, J.C.
Evans and W.W. Muelder

1,3,5-Tribromobenzene-D_2 ($C_6HD_2Br_3$)

Vibrational Analysis — Spectrochim. Acta
18, 1579 (1962) — J.R. Scherer, J.C.
Evans and W.W. Muelder

1,3,5-Tribromobenzene-D_3 ($C_6D_3Br_3$)

Vibrational Analysis — Spectrochim. Acta
18, 1579 (1962) — J.R. Scherer, J.C.
Evans and W.W. Muelder

2,4,6-Tribromophenol ($C_6H_2OHBr_3$)

Vibrational Analysis — Indian J. Phys. 48,
1089 (1974) — P.K. Mallik

1,2,3-Trichlorobenzene ($C_6H_3Cl_3$)

UV Absorption in Solution — Z. Phys. Chem.
B19, 76 (1932) — H. Conrad-Billroth
UV Absorption in Vapour — Chem. Rev. 41,
281 (1947) — H. Sponer
UV Absorption in Vapour — J. Opt. Soc. Amer.
39, 75 (1949) — H. Kohn and H. Sponer
Vibrational Analysis — J. Phys. Chem. 61,
730 (1957) — S.H. Hastings and D.E.
Nicholson
Vibrational Analysis — Spectrochim. Acta
19, 1739 (1963) — J.R. Scherer and
J.C. Evans
UV Absorption in Different States — Indian
J. Phys. 37, 173 (1963) — T.N. Misra
Calculation — Spectrochim. Acta 23A, 1489
(1967) — J.R. Scherer
Vibrational Analysis — Spectrochim. Acta
27A, 793 (1971) — J.H.S. Green, D.J.
Harrison and W. Kynaston

1,2,3-Trichlorobenzene-D_3 ($C_6D_3Cl_3$)

Vibrational Analysis — Spectrochim. Acta
19, 1739 (1963) — J.R. Scherer and
J.C. Evans

1,2,4-Trichlorobenzene ($C_6H_3Cl_3$)

UV Absorption in Vapour — Chem. Rev. 41,
281 (1947) — H. Sponer
UV Absorption in Vapour — J. Opt. Soc. Amer.
39, 75 (1949) — H. Kohn and H. Sponer
UV Absorption in Different States — Indian
J. Phys. 31, 483 (1957) — S.B. Banerjee
Vibrational Analysis — Indian J. Phys. 34,
554 (1960) — K.K. Deb and S.B. Banerjee
Fluorescence — Indian J. Phys. 34, 554
(1960) — K.K. Deb and S.B. Banerjee

1,2,4-Trichlorobenzene-D_3 ($C_6D_3Cl_3$)

Vibrational Analysis — Spectrochim. Acta
28A, 2233 (1972) — C.K. Ramamurty and
N.A. Narasimhan

1,3,5-Trichlorobenzene ($C_6H_3Cl_3$)

UV Absorption in Vapour — Chem. Rev. 41, 281 (1947) — H. Sponer
UV Absorption in Vapour — J. Opt. Soc. Amer. 39, 75 (1949) — H. Kohn and H. Sponer
IR Spectrum — Trans. Faraday Soc. 46, 103 (1950) — A.R.H. Cole and H.W. Thompson
UV Absorption in Different States — Indian J. Phys. 31, 483 (1957) — S.B. Banerjee
UV Absorption in Different States — Indian J. Phys. 31, 588 (1957) — S.B. Roy
IR Spectrum — Bull. Chem. Soc. Jap. 33, 1024 (1960) — S. Saeki
Vibrational Analysis — Spectrochim. Acta 18, 57 (1962) — J.R. Scherer, J.C. Evans, W.W. Muelder and J. Overend
IR Spectrum — Spectrochim. Acta 19, 844 (1963) — E.F. Mooney
Vibrational Analysis — Spectrochim. Acta 27A, 793 (1971) — J.H.S. Green, D.J. Harrison and W. Kynaston
Vibrational Analysis — Spectrochim. Acta 27A, 405 (1971) — J.E. Muller, T. Inoue, R.H. Laskin and H.D. Stidham

1,3,5-Trichlorobenzene-D_3 ($C_6D_3Cl_3$)

Vibrational Analysis — Spectrochim. Acta 18, 57 (1962) — J.R. Scherer, J.C. Evans, W.W. Muelder and J. Overend

2,3,5-Trichloro-p-Benzoquinone ($C_6HCl_3O_2$)

IR Spectrum — Indian J. Pure Appl. Phys. 9, 858 (1971) — G.D. Baruah, B.B. Lal and P.S. Dube

2,3,5-Trichloro-p-Benzoquinone-D_1 ($C_6DCl_3O_2$)

Vibrational Analysis — Spectrochim. Acta 34A, 453 (1978) — A. Girlando and C. Pecile

1,2,4-Trichloro-5-Fluorobenzene ($C_6H_2FCl_3$)

Vibrational Analysis — Spectrochim. Acta 26A, 849 (1970) — R.A. Nyquist

2,3,4-Trichlorophenol ($C_6H_2OHCl_3$)

IR Spectrum — Indian J. Pure Appl. Phys. 12, 529 (1974) — G.N.R. Tripathi and S. Ram

2,3,5-Trichlorophenol ($C_6H_2OHCl_3$)

UV Absorption in Vapour — Curr. Sci. 43, 713 (1974) — S.M. Pandey and S.J. Singh
Vibrational Analysis — Indian J. Phys. 48, 961 (1974) — S.M. Pandey and S.J. Singh

2,3,6-Trichlorophenol ($C_6H_2OHCl_3$)

IR Spectrum — Indian J. Pure Appl. Phys. 12, 529 (1974) — G.N.R. Tripathi and S. Ram

2,4,5-Trichlorophenol ($C_6H_2OHCl_3$)

UV Absorption in Vapour — Curr. Sci. 43, 713 (1974) — S.M. Pandey and S.J. Singh
Vibrational Analysis — Indian J. Phys. 48, 961 (1974) — S.M. Pandey and S.J. Singh

2,4,6-Trichlorophenol ($C_6H_2OHCl_3$)

UV Absorption in Different States — Indian J. Phys. 30, 553 (1956) — S.K. Sen
Vibrational Analysis — Indian J. Phys. 48, 961 (1974) — S.M. Pandey and S.J. Singh
Vibrational Analysis — Indian J. Phys. 48, 1089 (1974) — P.K. Mallik

2,4,6-Trichloropyridine ($C_5H_2NCl_3$)

Vibrational Analysis — Spectrochim. Acta 29A, 1177 (1973) — J.H.S. Green, D.J. Harrison and M.R. Kipps

1,3,5-Trichloropyrimidine ($C_4HN_2Cl_3$)

Vibrational Analysis — Spectrochim. Acta 25A, 219 (1969) — R.T. Bailey and D. Steele

1,3,5-Trichloropyrimidine-D_1 ($C_4DN_2Cl_3$)

Vibrational Analysis — Spectrochim. Acta 25A, 219 (1969) — R.T. Bailey and D. Steele

α-α-α-Trichlorotoluene ($C_6H_5CCl_3$)

UV Absorption Spectrum — J. Chem. Phys. 47,
2916 (1967) — K. Kimura and S. Nagakura
UV Absorption in Vapour — Appl. Spectrosc.
23, 549 (1969) — K.P.R. Nair, R. Amni
Amma and M.P. Srivastava

1,3,5-Trichlorotrifluorobenzene ($C_6Cl_3F_3$)

Vibrational Analysis — J. Mol. Spectrosc.
17, 334 (1965) — J.R. Nielsen and H.D.
Brandt
Vibrational Analysis — J. Mol. Spectrosc.
62, 228 (1976) — J.H.S. Green and
D.J. Harrison
Vibrational Analysis — J. Chem. Thermody-
namics 8, 529 (1976) — J.H.S. Green
and D.J. Harrison

1,3,5-Triethylbenzene ($C_6H_3(C_2H_5)_3$)

Vibrational Analysis — Indian J. Phys. 49,
873 (1975) — P.K. Mallik, S. Chakra-
vorty and S.B. Banerjee

1,2,4-Trifluorobenzene ($C_6H_3F_3$)

Vibrational Analysis — J. Chem. Phys. 21,
1457 (1953) — D.C. Smith, E.E. Ferguson,
R.L. Hudson and J.R. Nielsen
Vibrational Analysis — J. Chem. Phys. 21,
1727 (1953) — D.C. Smith, E.E. Ferguson,
R.L. Hudson and J.R. Nielsen
UV Absorption in Vapour — Can. J. Phys.
35, 322 (1957) — K.N. Rao and H. Sponer
Fluorescence — Can. J. Chem. 47, 597
(1969) — G.P. Semeluk, R.D.S. Stevens
and I. Unger
Vibrational Analysis — Spectrochim. Acta
27A, 807 (1971) — J.H.S. Green, D.J.
Harrison and W. Kynaston
Calculation — J. Mol. Spectrosc. 48, 446
(1973) — V.J. Eaton and D. Steele
Calculation — J. Mol. Struct. 32, 93
(1976) — F. Torok, A. Hegedus, K. Kosa
and P. Puley

1,3,5-Trifluorobenzene ($C_6H_3F_3$)

Vibrational Analysis — Disc. Faraday Soc.
9, 177 (1950) — J.R. Nielsen, C. Yu.
Liang and D.C. Smith
Vibrational Analysis — J. Chem. Phys. 21,
1727 (1953) — E.E. Ferguson, R.L.
Hudson, J.R. Nielsen and D.C. Smith
Calculation — J. Chem. Phys. 21, 886
(1953) — E.E. Ferguson
Calculation — Spectrochim. Acta 17, 1049
(1961) — G. Nonnenmacher and R. Mecke
Vibrational Analysis — Spectrochim. Acta
18, 1579 (1962) — J.R. Scherer, J.C.
Evans and W.W. Muelder
Vibrational Analysis — Spectrochim. Acta
19, 601 (1963) — J.R. Scherer
UV Absorption in Vapour — J. Chem. Phys.
39, 1253 (1963) — S.H. Bauer and C.F.
Alen
Fluorescence — Can. J. Chem. 42, 597
(1969) — G.P. Smeluk, R.D.S. Stevens
and I. Unger
Vibrational Analysis — Spectrochim. Acta
27A, 793 (1971) — J.H.S. Green, D.J.
Harrison and W. Kynaston
Calculation — J. Mol. Spectrosc. 48, 446
(1973) — V.J. Eaton and D. Steele
Calculation — J. Mol. Struct. 32, 93
(1976) — F. Torok, A. Hegedus, K. Kosa
and P. Puley

1,3,5-Trifluorobenzene-D_1 ($C_6H_2F_3D$)

Vibrational Analysis — Spectrochim. Acta
18, 1579 (1962) — J.R. Scherer, J.C.
Evans and W.W. Muelder

1,3,5-Trifluorobenzene-D_2 ($C_6HF_3D_2$)

Vibrational Analysis — Spectrochim. Acta
18, 1579 (1962) — J.R. Scherer, J.C.
Evans and W.W. Muelder

1,3,5-Trifluorobenzene-D_3 ($C_6F_3D_3$)

Vibrational Analysis — Spectrochim. Acta
18, 1579 (1962) — J.R. Scherer, J.C.
Evans and W.W. Muelder

o-Trifluoromethylbenzonitrile ($C_6HF_3CH_3CN$)

UV Absorption Spectrum — Curr. Sci. 46,
301 (1977) — M.R. Padhye and V.
Raghavendra
Fluorescence — Curr. Sci. 46, 301 (1977)—
M.R. Padhye and V. Raghavendra

m-Trifluoromethylbenzonitrile ($C_6HF_3CH_3CN$)

UV Absorption Spectrum — Curr. Sci. <u>46</u>, 301 (1977) — M.R. Padhye and V. Raghavendra

Fluorescence — Curr. Sci. <u>46</u>, 301 (1977) — M.R. Padhye and V. Raghavendra

p-Trifluoromethylbenzonitrile ($C_6HF_3CH_3CN$)

UV Absorption Spectrum — Curr. Sci. <u>46</u>, 301 (1977) — M.R. Padhye and V. Raghavendra

Fluorescence — Curr. Sci. <u>46</u>, 301 (1977) — M.R. Padhye and V. Raghavendra

2,4,6-Trifluoropyridine ($C_5H_2F_3N$)

Vibrational Analysis — Spectrochim. Acta <u>29A</u>, 1177 (1973) — J.H.S. Green, D.J. Harrison and M.R. Kipps

2,4,6-Trifluoropyrimidine ($C_4HF_3N_2$)

Vibrational Analysis — Spectrochim. Acta <u>23A</u>, 2989 (1967) — R.T. Bailey and D. Steele

Vibrational Analysis — Spectrochim. Acta <u>29A</u>, 1177 (1973) — J.H.S. Green, D.J. Harrison and M.R. Kipps

α−β− β −Trifluorostyrene($C_6H_5CF{:}CF_2$)

Vibrational Analysis — Spectrochim. Acta <u>33A</u>, 723 (1977) — M. Towland, A.M. North and R.A. Pethrick

1,3,5-Trifluorotrichlorobenzene ($C_6F_3Cl_3$)

IR Spectrum — J. Mol. Spectrosc. <u>17</u>, 334 (1965) — J.R. Nielsen and H.D. Brandt

1,2,3-Trihydroxybenzene ($C_6H_3(OH)_3$)

UV Absorption in Different States — Indian J. Phys. <u>37</u>, 173 (1963) — T.N. Misra

UV Absorption in Solution — J. Amer. Chem. Soc. <u>78</u>, 3445 (1957) — L. Jurd

Calculation — J. Mol. Struct. <u>6</u>, 246 (1970) — P.C. Mishra and D.K. Rai

IR Spectrum — J. Sci. Res. BHU (India) — <u>21</u>, 239 (1970) — V.N. Verma

1,2,4-Trihydroxybenzene ($C_6H_3(OH)_3$)

Calculation — J. Mol. Struct. <u>6</u>, 246 (1970) — P.C. Mishra and D.K. Rai

1,3,5-Trihydroxybenzene ($C_6H_3(OH)_3$)

Calculation — Spectrochim. Acta <u>17</u>, 1059 (1961) — G. Nonnenmacher and R. Mecke

Calculation — J. Mol. Struct. <u>6</u>, 246 (1970) — P.C. Mishra and D.K. Rai

Vibrational Analysis — Indian J. Phys. <u>49</u>, 873 (1975) — P.K. Mallik, S. Chakravorti and S.B. Banerjee

1,2,3-Trimethoxybenzene ($C_6H_3(OCH_3)_3$)

Vibrational Analysis — Anal. Chem. <u>13</u>, 700 (1947) — M.R. Fenske, W.G. Braun, R.V. Wiegand, D. Quiggle, R.M. McCormick and D.H. Rank

Vibrational Analysis — Opt. Spektrosk. <u>13</u>, 331 (1962) — A.M. Bogomolov

Vibrational Analysis — Spectrochim. Acta <u>27A</u>, 793 (1971) — J.H.S. Green, D.J. Harrison and W. Kynaston

Vibrational Analysis — Indian J. Phys. <u>51B</u>, 71 (1977) — A.K. Sarkar, S. Chakravorti and S.B. Banerjee

1,2,4-Trimethoxybenzene ($C_6H_3(OCH_3)_3$)

Vibrational Analysis — Indian J. Phys. <u>51B</u>, 71 (1977) — A.K. Sarkar, S. Chakravorti and S.B. Banerjee

1,3,5-Trimethoxybenzene ($C_6H_3(OCH_3)_3$)

Vibrational Analysis — Indian J. Phys. <u>51B</u>, 71 (1977) — A.K. Sarkar, S. Chakravorti and S.B. Banerjee

1,2,3-Trimethylbenzene ($C_6H_3(CH_3)_3$)

Raman — Monatsh. Chem. <u>65</u>, 13 (1934) — K.W.F. Kohlrausch and A. Pongratz

IR Spectrum — Disc. Faraday Soc. <u>9</u>, 100 (1950) — E.K. Plyer

IR Spectrum – J. Phys. Chem. $\underline{61}$, 730 (1957) – S.H. Hastings and D.E. Nicholson

Calculation – Opt. Spektrosk. $\underline{13}$, 183 (1962) – A.M. Bogomolov

UV Absorption in Vapour – Bull. Chem. Soc. Jap. $\underline{44}$, 2031 (1971) – V.N. Verma, K.P.R. Nair and D.K. Rai

Vibrational Analysis – Spectrochim. Acta $\underline{27A}$, 793 (1971) – J.H.S. Green, D.J. Harrison and W. Kynaston

Fluorescence – Chem. Phys. Lett. $\underline{14}$, 404 (1972) – G.M. Breuer and E.K.C. Lee

Vibrational Analysis – Spectrosc. Lett. $\underline{8}$, 349 (1975) – V.N. Verma

1,2,4-Trimethylbenzene $(C_6H_3(CH_3)_3)$

Raman – Monatsh. Chem. $\underline{65}$, 13 (1934) – K.W.F. Kohlrausch and A. Pongratz

Vibrational Analysis – Anal. Chem. $\underline{19}$, 700 (1947) – M.R. Fenske, W.G. Braun, R.V. Wiegand, D. Quiggle, R.M. McCormick and D.H. Rank

IR Spectrum – Disc. Faraday Soc. $\underline{9}$, 100 (1950) – E.K. Plyer

UV Absorption in Vapour – Proc. Natl. Inst. Sci. (India) $\underline{17}$, 385 (1951) – K. Sreeramamurty

Vibrational Analysis – Izv. Akad. Nauk. SSSR $\underline{19}$, 225 (1955) – V.T. Aleksanian, Kh.E. Sterin, A.L. Liebermann, E.A. Mikhailava, M.A. Prianishnikova and B.A. Kazanski

IR Spectrum – J. Phys. Chem. $\underline{61}$, 730 (1957) – S.H. Hastings and D.E. Nicholson

Raman – Indian J. Phys. $\underline{34}$, 554 (1960) – K.K. Deb and S.B. Banerjee

IR Spectrum – Indian J. Phys. $\underline{34}$, 1 (1960) – S.B. Banerjee and K.C. Medhi

Vibrational Analysis – Spectrochim. Acta $\underline{17}$, 1049 (1961) – G. Nonnenmacher and R. Mecke

Calculation – Opt. Spektrosk. $\underline{13}$, 331 (1962) – A.M. Bogomolov

Vibrational Analysis – Indian J. Phys. $\underline{37}$, 45 (1963) – K.K. Deb

UV Absorption in Different States – Indian J. Phys. $\underline{37}$, 299 (1963) – T.N. Misra

Vibrational Analysis – Spectrochim. Acta $\underline{27A}$, 807 (1971) – J.H.S. Green, D.J. Harrison and W. Kynaston

Vibrational Analysis – Spectrochim. Acta $\underline{28A}$, 33 (1972) – J.H.S. Green, D.J. Harrison and W. Kynaston

Fluorescence – Chem. Phys. Lett. $\underline{14}$, 404 (1972) – G.M. Breuer and E.K.C. Lee

Vibrational Analysis – Spectrosc. Lett. $\underline{8}$, 349 (1975) – V.N. Verma

1,3,5-Trimethylbenzene $(C_6H_3(CH_3)_3)$

UV Absorption in Solution – Z. Phys. Chem. $\underline{B19}$, 76 (1932) – H. Conrad-Billroth

UV Absorption in Solution – Z. Phys. Chem. $\underline{B21}$, 389 (1933) – K.L. Wolf and O. Strasser

Raman – Philos. Mag. 17th Series $\underline{15}$, 263 (1933) – S. Venkateshwaran

Raman – Monatsh. Chem. $\underline{74}$, 175 (1942) – E. Herz and K.W.F. Kohlrausch

Vibrational Analysis – J. Amer. Chem. Soc. $\underline{65}$, 803 (1943) – K.S. Pitzer and D.W. Scott

UV Absorption in Vapour – Che. Rev. $\underline{41}$, 281 (1947) – H. Sponer

Vibrational Analysis – Anal. Chem. $\underline{19}$, 700 (1947) – M.R. Fenske, W.G. Braun, R.V. Wiegand, D. Quiggle, R.M. McCormick and D.H. Rank

Fluorescence – Zh. Eksp. Teor. Fiz. $\underline{19}$, 1000 (1949) – P.P. Dirum and B.I. Sveshnikov

IR Spectrum – Disc. Faraday Soc. $\underline{9}$, 100 (1950) – E.K. Plyer

Calculation – Proc. Natl. Inst. Sci. (India) $\underline{20}$, 576 (1954) – V. Santhamma

Calculation – Curr. Sci. $\underline{23}$, 118 (1954) – V. Santhamma

Vibrational Analysis – Spectrochim. Acta $\underline{7}$, 253 (1955) – D.H. Whiffen

Calculation – Opt. Spektrosk. $\underline{1}$, 348 (1956) – M.A. Kovner

UV Absorption in Solution – Indian J. Phys. $\underline{31}$, 588 (1957) – S.B. Roy

UV Absorption in Different States – Indian J. Phys. $\underline{33}$, 41 (1959) – S.K. Sen

IR Spectrum – Indian J. Phys. $\underline{34}$, 1 (1960) – S.B. Banerjee and K.C. Medhi

Calculation – Spectrochim. Acta $\underline{17}$, 1049 (1961) – G. Nonnenmacher and R. Mecke

Vibrational Analysis – Nature $\underline{195}$, 595 (1962) – J.H.S. Green, W. Kynaston and H.A. Gebbie

Calculation – Opt. Spektrosk. $\underline{13}$, 331 (1962) – A.M. Bogomolov

Normal Coordinate Analysis – Spectrochim. Acta $\underline{18}$, 57 (1962) – J.R. Scherer, J.C. Evans, W.W. Muelder and J. Overend

Vibrational Analysis – Spectrochim. Acta $\underline{19}$, 807 (1963) – J.H.S. Green, W. Kynaston and H.A. Gebbie

Fluorescence – Proc. Roy. Soc. A283, 83
(1965) – J.B. Birks, C.L. Braga and
M.D. Lumb
UV Absorption in Vapour – Bull. Chem. Soc.
Jap. 44, 2031 (1971) – V.N. Verma,
K.P.R. Nair and D.K. Rai
Vibrational Analysis – Spectrochim. Acta
27A, 2073 (1971) – C. La Lau and R.G.
Snyder
Vibrational Analysis – Spectrochim. Acta
27A, 793 (1971) – J.H.S. Green, D.J.
Harrison and W. Kynaston
Vibrational Analysis – Spectrochim. Acta
28A, 33 (1972) – J.H.S. Green, D.J.
Harrison and W. Kynaston
UV Absorption in Vapour – Spectrosc. Lett.
6, 591 (1973) – V.N. Verma
Vibrational Analysis – Spectrosc. Lett. 8,
349 (1975) – V.N. Verma

2,3,4-Trimethylpyridine $(C_5H_2N(CH_3)_3)$

Vibrational Analysis – Chem. Pharm. Bull.
(Tokyo) 1, 146 (1953) – K. Tsuda and
M. Maruyama
Vibrational Analysis – Chem. Pharm. Bull.
(Tokyo) 4, 192 (1956) – H. Sindo and
N. Ikekawa
Vibrational Analysis – Spectrochim. Acta
29A, 293 (1973) – J.H.S. Green and
D.J. Harrison

2,3,5-Trimethylpyridine $(C_5H_2N(CH_3)_3)$

Vibrational Analysis – Chem. Pharm. Bull.
(Tokyo) 1, 146 (1953) – K. Tsuda and
M. Maruyama
Vibrational Analysis – Chem. Pharm. Bull.
(Tokyo) 4, 192 (1956) – H. Sindo and
N. Ikekawa
Vibrational Analysis – Spectrochim. Acta
29A, 293 (1973) – J.H.S. Green and
D.J. Harrison

2,3,6-Trimethylpyridine $(C_5H_2N(CH_3)_3)$

Vibrational Analysis – Chem. Pharm. Bull.
(Tokyo) 1, 146 (1953) – K. Tsuda and
M. Maruyama
Vibrational Analysis – Chem. Pharm. Bull.
(Tokyo) 4, 192 (1956) – H. Sindo and N.
Ikekawa
Vibrational Analysis – Spectrochim. Acta
29A, 293 (1973) – J.H.S. Green and D.J.
Harrison

2,4,5-Trimethylpyridine $(C_5H_2N(CH_3)_3)$

Vibrational Analysis – Chem. Pharm. Bull.
(Tokyo) 1, 283 (1953) – K. Tsuda and
M. Maruyama
Vibrational Analysis – Chem. Pharm. Bull.
(Tokyo) 4, 142 (1956) – H. Sindo and
N. Ikekawa
Vibrational Analysis – Spectrochim. Acta
29A, 293 (1973) – J.H.S. Green and
D.J. Harrison

2,4,6-Trimethylpyridine $(C_5H_2N(CH_3)_3)$

UV Absorption in Solution – J. Chem. Soc.
692 (1910) – J.E. Purvis
Raman – Z. Phys. Chem. B53, 124 (1943) – E.
Herz, L. Kahovec and K.W.F. Kohlrausch
Vibrational Analysis – Chem. Pharm. Bull.
(tokyo) 1, 146 (1953) – K. Tsuda and
M. Maruyama
Vibrational Analysis – Chem. Pharm. Bull.
(Tokyo) 4, 192 (1956) – H. Sindo and N.
Ikekawa
UV Absorption in Different States – Indian
J. Phys. 45, 115 (1971) – S.C. Bag
Vibrational Analysis – Spectrochim. Acta
29A, 293 (1973) – J.H.S. Green and
D.J. Harrison

2,4,6-Trinitroaniline $(C_6H_2NH_2(NO_2)_3)$

IR Spectrum – Aust. J. Chem. 11, 513 (1958) –
L.K. Dyall and A.N. Hambly
IR Spectrum – Spectrochim. Acta 17, 291
(1961) – L.K. Dyall

1,3,5-Trinitrobenzene $(C_6H_3(NO_2)_3)$

Vibrational Analysis – C.R. Acad. Bulg. Sci.
17, 113 (1964) – P. Simova, M. Popava,
Kh. Dimitrov and N. Petzev
Vibrational Analysis – Can. J. Chem. 47,
2515 (1969) – H.F. Shurvell, A.R. Norris
and D.E. Irish

1,3,5-Trinitrobenzene-D₃ $(C_6D_3(NO_2)_3)$

Vibrational Analysis – C.R. Acad. Bulg. Sci.
17, 113 (1964) – P. Simova, M. Popava,
Kh. Dimitrov and N. Petzev
Vibrational Analysis – Can. J. Chem. 47,
2515 (1969) – H.F. Shurvell, A.R. Norris
and D.E. Irish

Veratrole $(C_6H_4(OCH_3)_2)$

See   o-Dimethoxybenzene

2-Vinylpyridine $(C_5H_4NCH_2{:}CH)$

Vibrational Analysis — Spectrochim. Acta
33A, 249 (1977) — J.H.S. Green and D.J.
Harrison

4-Vinylpyridine $(C_5H_4NCH_2{:}CH)$

Vibrational Analysis — Spectrochim. Acta
33A, 249 (1977) — J.H.S. Green and
D.J. Harrison

o-Xylene $(C_6H_4(CH_3)_2)$

Raman — Monatsh. Chem. 74, 160 (1943) —
E. Herz
Vibrational Analysis — J. Amer. Chem. Soc.
65, 803 (1943) — K.S. Pitzer and D.W.
Scott
Vibrational Analysis — Anal. Chem. 19, 700
(1947) — M.R. Fenske, W.G. Braun, R.V.
Wiegand, R.M. McCormick and D.H. Rank
Vibrational Analysis — Izv. Akad. Nauk.
SSSR 501 (1950) — P.A. Bazhulin, S.A.
Ukholin, A.L. Liebermann, S.S. Novikov
and B.A. Kazanski
UV Absorption in Vapour — Disc. Faraday Soc.
9, 53 (1950) — W.C. Price, V.J. Hammond,
J.P. Teegan and A.D. Walsh
UV Absorption in Vapour — J. Chem. Phys.
20, 1248 (1952) — C.D. Cooper and H.
Sponer
UV Absorption in Different States — Indian
J. Phys. 26, 233 (1952) — H.N. Swamy
UV Absorption in Vapour — J. Sci.Res. BHU
(India) 7, 171 (1957) — I.S. Singh
IR Spectrum — J. Phys. Chem. 61, 730 (1957)—
S.H. Hastings and D.E. Nicholson
Vibrational Analysis — Opt. Spektrosk. 7,
751 (1959) — M.A. Kovner and A.M.
Bogomolov
Calculation — Opt. Spektrosk. 10, 322
(1961) — A.M. Bogomolov
Calculation — Spectrochim. Acta 17, 1049
(1961) — G. Nonnenmacher and R. Mecke
Vibrational Analysis — Spectrochim. Acta
18, 1433 (1962) — K. Venkateshwarlu
and M. Radhakrishnan
Raman — Opt. Spektrosk. 13, 783 (1962) —
N.K. Sidorov, L.S. Stal'makhova and L.I.
Bratamova

IR Spectrum — Spectrochim. Acta 19, 807
(1963) — J.H.S. Green, W. Kynaston and
H.A. Gebbie
Fluorescence — J. Mol. Spectrosc. 23, 365
(1967) — M.D. Lumb and D.A. Weyl
IR Spectrum — J. Chem. Phys. 52, 6426
(1970) — G.W.F. Pardoe, S.J. Larsen
and H.A. Gebbie
Vibrational Analysis — Spectrochim. Acta
26A, 1913 (1970) — J.H.S. Green
Raman — J. Mol. Struct. 5, 477 (1970) —
N.T. McDevitt and W.G. Fateley
UV Absorption in Vapour — J. Phys. Chem.
75, 2741 (1971) — W.A. Noyes,Jr and
D.A. Harter
UV Emission Spectrum — Chem. Lett, 1157
(1972) — T. Ogawa, M. Tsuji, M. Toyoda
and N. Ishibashi
Fluorescence — Chem. Phys. Lett. 14, 404
(1972) — G.M. Breuer and E.K.C. Lee

m-Xylene $(C_6H_4(CH_3)_2)$

Vibrational Analysis — J. Amer. Chem. Soc.
65, 803 (1943) — K.S. Pitzer and D.W.
Scott
Raman — Monatsh. Chem. 74, 160 (1943) —
E. Herz
UV Absorption in Vapour — Disc. Faraday
Soc. 9, 53 (1950) — W.C. Price, V.J.
Hammond, J.P. Teegan and A.D. Walsh
UV Absorption in Vapour — J. Chem. Phys.
20, 1248 (1952) — C.D. Cooper and H.
Sponer
Raman — Indian J. Phys. 28, 307 (1954) —
D.C. Biswas
Calculation — J. Chem. Phys. 23, 1997
(1955) — R.S. Mulliken
Vibrational Analysis — Can. J. Chem. 35,
911 (1957) — J.K. Wilmshurt and H.J.
Bernstein
UV Absorption in Vapour — J. Sci. Res. BHU
(India) — 8, 19 (1957) — I.S. Singh
Calculation — Opt. Spektrosk. 4, 301 (1958)—
M.A. Kovner and A.M. Bogomolov
Calculation — Spectrochim. Acta 17, 1049
(1961) — G. Nonnenmacher and R. Mecke
Raman — Opt. Spektrosk. 13, 783 (1962) —
N.K. Sidorov, L.S. Stal'makhova and L.I.
Bratomova
Vibrational Analysis — Opt. Spektrosk. 13,
159 (1962) — A.M. Bogomolov
Vibrational Analysis — Spectrochim. Acta
20, 1343 (1964) — E.F. Mooney
Vibrational Analysis — J. Chim. Phys. 63,
552 (1966) — C. Garrigou-Lagrange, M.
Chehata and J. Lascombe

Fluorescence – J. Mol. Spectrosc. 23, 365
(1967) – M.D. Lumb and D.A. Weyl
Vibrational Analysis – J. Mol. Spectrosc.
32, 247 (1969) – Y. Kakiuti, H. Saito
and T. Yokoyama
Vibrational Analysis – Spectrochim. Acta
26A, 1523 (1970) – J.H.S. Green
Raman – J. Mol. Struct. 5, 477 (1970) –
N.T. McDevitt and W.G. Fateley
Vibrational Analysis – Spectrochim. Acta
27A, 2073 (1971) – C. La Lau and R.G.
Snyder
UV Absorption Spectrum – J. Phys. Chem. 75,
2741 (1971) – W.A. Noyes, Jr and D.A.
Harter
Fluorescence – Chem. Phys. Lett. 14, 404
(1972) – G.M. Breuer and E.K.C. Lee

m-Xylene-α-α'-D_6 ($C_6H_4(CD_3)_2$)

Vibrational Analysis – Can. J. Chem. 35,
911 (1957) – J.K. Wilmshurt and H.J.
Bernstein
Vibrational Analysis – Spectrochim. Acta
27A, 2073 (1971) – C. La Lau and R.G.
Snyder

p-Xylene ($C_6H_4(CH_3)_2$)

Vibrational Analysis – J. Amer. Chem. Soc.
65, 803 (1943) – K.S. Pitzer and D.W.
Scott
UV Absorption in Vapour – Disc. Faraday
Soc. 9, 53 (1950) – W.C. Price, V.J.
Hammond, J.P. Teegan and A.D. Walsh
Vibrational Analysis – Izv. Akad. Nauk SSSR
501 (1950) – P.A. Bazhulin, S.A. Ukholin,
A.L. Liebermann, S.S. Novikov and B.A.
Kazanski
UV Absorption in Vapour – J. Chem. Phys. 20,
607 (1952) – C.D. Cooper and M.L.N.
Sastri
Fluorescence – J. Chem. Phys. 20, 607
(1952) – C.D. Cooper and M.L.N. Sastri
UV Absorption in Different States – Indian
J. Phys. 26, 233 (1952) – H.N. Swamy
UV Absorption in Vapour – J. Chem. Phys. 20,
1248 (1952) – C.D. Cooper and H. Sponer
Raman – Indian J. Phys. 28, 307 (1954) –
D.C. Biswas
Calculation – Dokl. Akad. Nauk SSSR 97, 229
(1954) – M.A. Kovner
UV Absorption in Vapour – J. Sci. Res. BHU
(India) 8, 32 (1957) – I.S. Singh
Vibrational Analysis – Spectrochim. Acta
12, 305 (1958) – C. Garrigou-Lagrange,
J.M. Lebas and M.L. Josien

Vibrational Analysis – Spectrochim. Acta
15, 225 (1959) – C. Garrigou-Lagrange,
J.M. Lebas and M.L. Josien
Calculation – Z. Elektrochem. 64, 940
(1960) – E.W. Schmid, J. Brandmuller
and G. Nonnenmacher
Vibrational Analysis – J. Chem. Phys. 37,
867 (1962) – D.W. Scott, J.F. Messerly,
S.S. Todd, I.A. Hossenlopp, D.R. Douslin
and J.P. McCullough
Vibrational Analysis – Opt. Spektrosk. 12,
186 (1962) – A.M. Bogomolov
Raman – Opt. Spektrosk. 13, 783 (1962) –
N.K. Sidorov, L.S. Stal'makhova and
L.I. Bratamova
Fluorescence – Proc. Roy. Soc. A283, 83
(1965) – J.B. Birks, C.L. Braga and
M.D. Lumb
Vibrational Analysis – Proc. Roy. Soc.
A298, 51 (1967) – P.R. Griffiths and
H.W. Thompson
Fluorescence – J. Mol. Spectrosc. 23, 365
(1967) – M.D. Lumb and D.A. Weyl
Vibrational Analysis – Spectrochim. Acta
26A, 1503 (1970) – J.H.S. Green
Raman – J. Mol. Struct. 5, 477 (1970) –
N.T. McDevitt and W.G. Fateley
Vibrational Analysis – Spectrochim. Acta
27A, 2073 (1971) – C.La Lau and R.G.
Snyder
UV Absorption Spectrum – J. Phys. Chem. 75,
2741 (1971) – W.A. Noyes, Jr and D.A.
Harter
Fluorescence – Bull. Chem. Soc. Jap. 46,
2637 (1973) – T. Ogawa, M. Tsuji, M.
Toyoda and N. Ishibashi

p-Xylene-α-α-α-D_3 ($C_6H_4CH_3CD_3$)

Vibrational Analysis – Spectrochim. Acta
27A, 1337 (1971) – S. Julien-Leferriere
and J.M. Lebas

p-Xylene-2,3,5,6-D_4 ($C_6D_4(CH_3)_2$)

Vibrational Analysis – Spectrochim. Acta
27A, 1337 (1971) – S. Julien-Leferriere
and J.M. Lebas

p-Xylene-α-α'-D_6 ($C_6H_4(CD_3)_2$)

Vibrational Analysis – Spectrochim. Acta
27A, 2073 (1971) – C. La Lau and R.G.
Snyder
Vibrational Analysis – Spectrochim. Acta
27A, 1337 (1971) – S. Julien-Leferriere
and J.M. Lebas

2,3-Xylenol $(C_6H_3OH(CH_3)_2)$

IR Spectrum — Indian J. Pure Appl. Phys. <u>10</u>, 489 (1972) — M.A. Shashidhar, K.S. Rao and E.S. Jayadeo
Vibrational Analysis — Spectrochim. Acta <u>28A</u>, 33 (1972) — J.H.S. Green, D.J. Harrison and W. Kynaston
Vibrational Analysis — Curr. Sci. <u>41</u>, 284 (1972) — B.B. Lal, G.D. Baruah and I.S. Singh
UV Absorption in Vapour — Indian J. Phys. <u>48</u>, 62 (1974) — D. Marjit and S.B. Banerjee
UV Absorption (Hydrogen Bonding) — Spectrochim. Acta <u>34A</u>, 617 (1978) — T. Ganguly and S.B. Banerjee

2,4-Xylenol $(C_6H_3OH(CH_3)_2)$

IR Spectrum — Indian J. Phys. <u>41</u>, 230 (1967) — S.B. Banerjee and D.K. Mukharjee
UV Absorption in Vapour — Indian J. Pure Appl. Phys. <u>8</u>, 239 (1970) — G.D. Baruah, R. Amni Amma and K.P.R. Nair
UV Absorption in Different States — Indian J. Phys. <u>44</u>, 339 (1970) — D. Marjit
Vibrational Analysis — Indian J. Pure Appl. Phys. <u>10</u>, 302 (1972) — G.D. Baruah, P.S. Dube, Sri Singh and S.R. Singh
Vibrational Analysis — Spectrochim. Acta <u>28A</u>, 33 (1972) — J.H.S. Green, D.J. Harrison and W. Kynaston

2,4-Xylenol-OD $(C_6H_3OD(CH_3)_2)$

Vibrational Analysis — Spectrochim. Acta <u>28A</u>, 33 (1972) — J.H.S. Green, D.J. Harrison and W. Kynaston

2,5-Xylenol $(C_6H_3OH(CH_3)_2)$

IR Spectrum — Indian J. Pure Appl. Phys. <u>10</u>, 489 (1972) — M.A. Shashidhar, K.S. Rao and E.S. Jayadeo
Vibrational Analysis — Spectrochim. Acta <u>28A</u>, 33 (1972) — J.H.S. Green, D.J. Harrison and W. Kynaston
IR Spectrum — Curr. Sci. <u>41</u>, 448 (1972) — B.B. Lal, G.D. Baruah and I.S. Singh
UV Absorption in Vapour — Indian J. Phys. <u>48</u>, 62 (1974) — D. Marjit and S.B. Banerjee
UV Absorption (Hydrogen Bonding) — Spectrochim. Acta <u>34A</u>, 617 (1978) — T. Ganguly and S.B. Banerjee

2,6-Xylenol $(C_6H_3OH(CH_3)_2)$

IR Spectrum — Indian J. Pure Appl. Phys. <u>10</u>, 489 (1972) — M.A. Shashidhar, K.S. Rao and E.S. Jayadeo
Vibrational Analysis — Spectrochim. Acta <u>28A</u>, 33 (1972) — J.H.S. Green, D.J. Harrison and W. Kynaston
Vibrational Analysis — Indian J. Pure Appl. Phys. <u>10</u>, 302 (1972) — G.D. Baruah, P.S. Dube, Sri Singh and S.R. Singh
UV Absorption in Vapour — Indian J. Phys. <u>49</u>, 329 (1975) — A.K. Sarkar and D. Marjit
UV Absorption (Hydrogen Bonding) — Spectrochim. Acta <u>34A</u>, 617 (1978) — T. Ganguly and S.B. Banerjee

3,4-Xylenol $(C_6H_3OH(CH_3)_2)$

IR Spectrum — Indian J. Pure Appl. Phys. <u>10</u>, 489 (1972) — M.A. Shashidhar, K.S. Rao and E.S. Jayadeo
Vibrational Analysis — Spectrochim. Acta <u>28A</u>, 33 (1972) — J.H.S. Green, D.J. Harrison and W. Kynaston

3,5-Xylenol $(C_6H_3OH(CH_3)_2)$

Raman — Monatsh. Chem. <u>76</u>, 213 (1947) — K.W.F. Kohlrausch
IR Spectrum — Indian J. Phys. <u>41</u>, 230 (1967) — S.B. Banerjee and D.K. Mukharjee
IR Spectrum — Indian J. Pure Appl. Phys. <u>10</u>, 489 (1972) — M.A. Shashidhar, K.S. Rao and E.S. Jayadeo
Vibrational Analysis — Curr. Sci. <u>41</u>, 284 (1972) — B.B. Lal, G.D. Baruah and I.S. Singh
Vibrational Analysis — Spectrochim. Acta <u>28A</u>, 33 (1972) — J.H.S. Green, D.J. Harrison and W. Kynaston
UV Absorption in Vapour — Indian J. Phys. <u>48</u>, 62 (1974) — D. Marjit and S.B. Banerjee
UV Absorption (Hydrogen Bonding) — Spectrochim. Acta <u>34A</u>, 617 (1978) — T. Ganguly and S.B. Banerjee

Xylidine $(C_6H_3NH_2(CH_3)_2)$

See   Dimethylaniline